STUDIES IN STATISTICAL MECHANICS

Volume V

Editors: J. DE BOER and G. E. UHLENBECK

The dispersion of sound in monoatomic gases

by James D. Foch, Jr. and George W. Ford

The kinetic theory of gases

by Mrs. C. S. Wang Chang and G. E. Uhlenbeck

NORTH-HOLLAND/AMERICAN ELSEVIER

STUDIES IN STATISTICAL MECHANICS

CONTENTS OF VOLUME I (1962)

CONTENTS OF VOLUME II (1964)

CONTENTS OF VOLUME III (1965)

CONTENTS OF VOLUME IV (1969)

STUDIES IN STATISTICAL MECHANICS

VOLUME V

EDITORS:

J. DE BOER

University of Amsterdam

and

G. E. UHLENBECK

The Rockefeller Institute, New York

1970

NORTH-HOLLAND PUBLISHING COMPANY – AMSTERDAM · LONDON
AMERICAN ELSEVIER PUBLISHING COMPANY, INC. – NEW YORK

North-Holland ISBN 0 7204 1105 X
American Elsevier ISBN 0 444 10008 3

PUBLISHERS:

NORTH-HOLLAND PUBLISHING COMPANY – AMSTERDAM
NORTH-HOLLAND PUBLISHING COMPANY, LTD. – LONDON

SOLE DISTRIBUTORS FOR THE U.S.A. AND CANADA:

AMERICAN ELSEVIER PUBLISHING COMPANY, INC.
52 VANDERBILT AVENUE
NEW YORK, N.Y. 10017

PRINTED IN THE NETHERLANDS

PREFACE TO VOLUME V

The first part of this volume of the Studies contains a reprint of five reports by Mrs. C. S. Wang Chang and G. E. Uhlenbeck on the kinetic theory of gases. It follows our policy to make available articles which are difficult to obtain, which have some historical interest, and for which there is still some demand. We refer to the foreword by one of us for comments on the contents and origin of these reports.

The second part of the volume is devoted to a study of the theory of the propagation of sound in mono-atomic gases. It combines part of the dissertation of Dr. J. Foch written in 1967 and the work by Dr. G.W. Ford on the so-called Kac models which was discussed briefly in the Lectures in Statistical Mechanics (Amer. Math.Soc., Providence, Rhode Island, 1963) by G. E. Uhlenbeck and G. W. Ford, but which was never published in detail. It seems to us that this contribution together with the reports of Mrs. Chang on the dispersion of sound present a fine survey of an important topic of the kinetic theory of gases. They lead the student from the first attempts to generalize the classical results of Laplace and Kirchhoff to the present status, including the still open problems. However, we want to emphasize that in accordance with our general policy the articles do not claim to be a complete review of the subject. The literature is quite extensive and several interesting contributions could not be discussed.

The Editors

Part A

The kinetic theory of gases

C. S. WANG CHANG and G. E. UHLENBECK
University of Michigan (1948)

FOREWORD

G. E. UHLENBECK

After the war Mrs. C. S. Wang Chang and I started to work on a number of problems in the kinetic theory of gases. We were supported by a grant from the Office of Naval Research and we published our work in the form of a series of reports which were printed by the Engineering Research Institute of the University of Michigan in Ann Arbor. Although these reports were widely distributed, they never appeared in any journal or book. In this volume of the Studies in Statistical Mechanics five of the early reports are reprinted as they appeared except for minor corrections. Although I occasionally still get requests for copies, one might well question the motive for trying to save these reports from oblivion. They are dated and because of the form in which they appeared they are here and there a bit repetitious. They may have some historical interest, but the main reason for reprinting them is my belief that they may still serve as an introduction for the student in the kinetic theory of gases. This theory still attracts a great deal of attention and is still far from being completely clarified. I also hope that this volume will show the special qualities of Dr. C. S. Wang Chang, with whom I was associated from her student days at the University of Michigan till she returned in 1956 with her husband and child to China where she joined the Institute of Physics of the Chinese Academy of Sciences in Peking. Without her persistence and dedication the work would never have been completed.

The first report (Chapter I) is based on a lecture I gave before the American Physical Society in 1947, and it contains a survey of the problems which would occupy us and some of our first results. In the next two reports (Chapters II and III) Mrs. Chang gave the details of the calculations for the dispersion of sound and for the thickness of shock waves as derived from the Navier-Stokes and higher order hydrodynamical equations. Her results for shock waves have still not been verified experimentally. They are of great interest since they depend (in contrast with the sound dispersion) not only on the linear terms but also on the new non-linear terms in the so-called Burnett equations. In Chapter IV we attempted to derive the disperion and absorption of high frequency sound directly from the linearized Boltzmann equation. This report is still in demand although it is superseded by the later work of J. Foch and G. W. Ford, which will appear in the next volume of the Studies, and where also the comparison with experiment will be discussed. Finally, Chapter 5 is devoted to a special problem, a generalization of a problem proposed by Rayleigh, which is not related to the problems mentioned in Chapter I, but which I believe still has some interest.

All other reports which Mrs. Chang and I wrote have to do with the transition of the very dilute or Knudsen gas regime to the normal density or Claudius gas regime. Their titles are:

Transport phenomena in very dilute gases: I (1949) and II (1950) The heat transport between two parallel plates as function of the Knudsen number (1953)

The Couette flow between two parallel plates as function of the Knudsen number (1954)

On the behavior of a gas near a wall; a problem of Kramers (1956)

Some copies are still available.

I have not tried to make the reports more up to date by giving references to more recent literature since this would be a major undertaking. I have added a few notes at the end of the volume to correct some of the more glaring omissions. For a recent account see, for instance, the book by Carlo Cercignani, Mathematical Methods in Kinetic Theory (Plenum Press, New York, 1969).

TABLE OF CONTENTS

Chapter I

ON THE TRANSPORT PHENOMENA IN RARIFIED GASES

C. S. WANG CHANG and G. E. UHLENBECK
University of Michigan (1948)

1. INTRODUCTION

1.1. *Historical summary*

The explanation of the laws of the viscosity, heat-conduction, and diffusion in gases of normal pressures is one of the first successes of the kinetic theory. As is well known, for instance, about one hundred years ago Clausius showed on the basis of the kinetic theory that in the case of a gas under ordinary pressures placed between two plates of different temperatures, the heat flow is proportional to the temperature gradient, that the proportionality factor is independent of the pressure, increases with the average temperature (about as $\sqrt{T}$) *and* has the correct order of magnitude. The further development of the theory of non-uniform gases has taken much longer, and has not penetrated into the usual textbooks. The problem was to explain on the basis of the molecular interaction laws (which depend of course also on the molecular *model*) the general Stokes-Navier equations for compressible media and to calculate at the same time rigorously the transport coefficients in terms of molecular quantities. This general problem has been treated according to two different methods. The first started from a paper by Maxwell (1879) and was brought to completion by the work of Chapman (1911-1916). The second started from a paper by Lorentz (1905), was generalized by Hilbert (1912) and was brought to a final form in the dissertation of Enskog in 1917. The results of the two methods are identical and at present the difference in the methods is perhaps more apparent than it seemed originally. In the present report the Lorentz-Hilbert-Enskog method will be followed, because it is perhaps a little clearer and it shows better the *limitations* of the theory, in which we are most interested [1].

1.2. *Limitations of the theory*

There are essentially two limitations:

1) The gas must not be *too dense*, so that triple and higher order collisions can be neglected, and only binary collisions need be considered. It seems possible to generalize the theory towards higher densities but the details have not been worked out. Presumably (it has never been proved strictly) the Stokes-Navier equations remain valid, but the transport coefficients will become dependent on pressure, so that for instance the viscosity coefficient, μ, will be developable in the form

$$\mu = \mu_0 + \mu_1 p + \mu_2 p^2 + \dots ,$$

analogous to the well-known development in virial coefficients of the equation of state [1*]*.

2) The gas must not be *too dilute*, so that the relative variation of any of the macroscopic quantities over a mean free path is very small. This is the limitation in which we are mainly interested at present. The opposite limiting case, the case where the mean free path is large compared to the macroscopic lengths is again well-known. It is the case of the Knudsen gas. There are striking differences with the phenomena under moderate pressures. For instance, in the Couette flow, the force does not depend any more on the velocity gradient, but only on the velocity difference, the force depends linearly on the pressure, etc.

* Starred notes refer to the notes of the editor at the end of the volume.

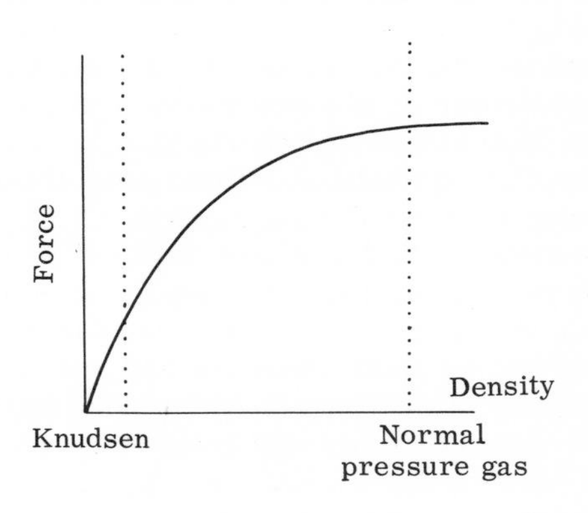

Fig. 1.

1.3. *The questions to be discussed*

The main question is what happens in the transition region. This is a very difficult question, and from a fundamental point of view not much is known about it. We will mainly be interested in the approach towards the normal pressure gas laws. It seems possible to discuss also the first deviations from the Knudsen gas laws (by considering a few molecular collisions), but the details have not been worked out. In this paper the following questions will be discussed.

1) What are the gas dynamical equations? It is almost certain that the Stokes-Navier equations will *not* be valid any more. How do they have to be generalized?

2) Will the state of the gas be determined (as in Stokes-Navier) by the five quantities: the density $\rho(x,y,z,t)$, the flow velocity $\boldsymbol{V}(x,y,z,t)$, and the temperature $T(x,y,z,t)$ or will more be needed?

3) What is the influence of the molecular structure (difference between monoatomic and polyatomic molecules)?

4) How must one describe the interaction between the gas and a solid? Are the classical boundary conditions still valid or must they be extended?

2. DIMENSIONAL CONSIDERATIONS

Before going into each of these theoretical questions it seems worthwhile to mention some dimensional considerations, which already give some insight and guidance in experimental investigations.

It is well known that at low speeds it follows from the Stokes-Navier equations that all dimensionless coefficients (of drag, lift, etc.) can only be functions of the Reynolds number, $\mathcal{R}$. For instance, for the flow through a circular tube the resistance coefficient C_{R} defined by

$$p_1 - p_2 = C_{\mathrm{R}} \frac{L}{R} \frac{\rho V^2}{2}$$

(where p_1 and p_2 are the pressures at the ends of the tube of length L, R is the radius of the tube, and ρ and V are the density of the gas and the average velocity of the flow respectively) is a function of

$$\mathcal{R} = \rho V R / \mu$$

only; and in fact for low speed (Poiseuille)

$$C_R = 16/\mathcal{R} \ .$$

There is also an old remark due to Reynolds which points out that Reynolds' number can be considered as the product of two dimensionless numbers. One has in fact from kinetic theory:

$$\mu \sim \rho \lambda c \ ,$$

where λ is the mean free path, c is an average thermal velocity. So

$$\mathcal{R} \sim \frac{V}{c} \cdot \frac{R}{\lambda} \ .$$

Only if $V/c \ll 1$ and $R/\lambda \gg 1$ will they enter into the theory as the product. In general (at high speeds, or low pressures) one may expect that both numbers will enter independent of each other so that then

$$C_R = F\left(\frac{V}{c}, \frac{R}{\lambda}\right)$$

Perhaps it is good to introduce new names:

$$K = \lambda/R = \text{Knudsen number}$$

(λ is now the Maxwell mean free path; for elastic spheres:

$$\lambda = 1/\sqrt{2}\pi n d^2 \ ,$$

where n is the number density and d is the diameter of the molecule), and

$$M_m = V/c = \text{Molecular Mach number} \ ,$$

where c is the root mean square molecular velocity (it is of the same order as the velocity of sound). Assuming that all dimensionless numbers depend only on K and M_m, one can try to find semi-empirical formula to represent the data. At least for small M_m one usually knows from the theory the dependence on K for very small and very large K, and one can then try to interpolate. For instance for the flow through tubes the experimental results of Knudsen can be represented by:

$$C_R = \frac{7.34K}{M_m\left[1 + 6.76K\frac{K+2.52}{K+3.11}\right]}$$

for a range of values of K between 10^{-2} and 10^3.

Of course, it should be emphasized that in such a program a definite *assumption* is involved, namely that in the interaction of a solid and a gas only the velocity and the dimensions of the solid are of importance. It seems unlikely that this assumption is strictly true; the structure of the solid-gas interface will no doubt play a more important role.

3. THE HILBERT-ENSKOG THEORY

To come back to the general questions, perhaps it is worthwhile to first give a short outline of the Hilbert-Enskog theory. The starting point of this theory is the Boltzmann equation:

$$\frac{\partial f}{\partial t} + \xi_\alpha \frac{\partial f}{\partial x_\alpha} + X_\alpha \frac{\partial f}{\partial \xi_\alpha} = \int \mathrm{d}\xi_1 \int \mathrm{d}\Omega\, g\, I(g,\theta)\,[f'f_1' - ff_1]\,, \quad (1)$$

which is the general continuity equation for the distribution function

$$f(x, y, z, \xi, \eta, \zeta, t)\, \mathrm{d}x\mathrm{d}y\mathrm{d}z\mathrm{d}\xi\, \mathrm{d}\eta\, \mathrm{d}\zeta\ .$$

Here, x, y, z, and ξ, η, ζ are the coordinates and velocity components of a molecule at the time t, X_i is the ith component of the outside force per unit mass acting on the molecule, f_1 is the distribution function corresponding to the velocity components ξ_1, η_1, ζ_1 and f' and f_1' are the distribution functions corresponding to the velocities after collision, $I(g,\theta)$ is the differential collision cross-section corresponding to the relative velocity g and the scattering angle θ. The determination of $I(g,\theta)$ requires a knowledge of the intermolecular force law. Equation (1) is valid only for *monoatomic* gases of such low density that only the *binary collisions* are of importance.

Strict consequences of the Boltzmann equation are still the general hydrodynamical equations:

$$\frac{\partial \rho}{\partial t} + \mathrm{div}\,(\rho \boldsymbol{u}) = 0\ ,$$

$$\rho \frac{\mathrm{D}u_i}{\mathrm{D}t} = \rho X_i - \frac{\partial P_{i\alpha}}{\partial x_\alpha}\ ,$$

$$\rho \frac{\mathrm{D}}{\mathrm{D}t}\left(\frac{Q}{\rho}\right) + \mathrm{div}\, \boldsymbol{q} = -P_{\alpha\beta} D_{\alpha\beta}\ , \quad (2)$$

where

$$\frac{\mathrm{D}}{\mathrm{D}t} \equiv \frac{\partial}{\partial t} + u_\alpha \frac{\partial}{\partial x_\alpha} ,$$

$$D_{ij} = \text{deformation tensor} = \frac{1}{2}\left(\frac{\partial u_i}{\partial x_j} + \frac{\partial u_j}{\partial x_i}\right) .$$

Equations (2) connect the average values of the density ρ, the velocity u_i, and the heat energy density Q (of the order $c_V T$) defined as follows:

$$\rho = m \int f \mathrm{d}\xi = mn ,$$

$$u_i = \frac{1}{n} \int \xi_i f \mathrm{d}\xi = \bar{\xi}_i ,$$

$$Q = \tfrac{1}{2}\rho \, \overline{u_\alpha u_\alpha} ,$$

where $u_i = \xi_i - u_i$. They express the general conservation laws of the number density, the impulse, and the energy. The equations contain further higher order average values:

$$P_{ij} = \text{stress tensor} = \rho \, \overline{u_i u_j}$$

and

$$q_i = \text{heat current density} = \tfrac{1}{2}\rho \, \overline{u_i u_\alpha u_\alpha} ,$$

which cannot yet be expressed in terms of ρ, $\boldsymbol{u}$, and T, so that equations (2) are still an empty frame.

Next one solves the Boltzmann equation by a successive approximation method. In the zeroth approximation

$$f = f^{(0)} = n\left(\frac{m}{2\pi kT}\right)^{\frac{3}{2}} \mathrm{e}^{-\frac{m}{2kT}(\xi_\alpha - u_\alpha)(\xi_\alpha - u_\alpha)} ,$$

in which the five quantities n, u_i, and T are still functions of x, y, z, and t. In other words, in the zeroth approximation there is local equilibrium. The next approximations involve space derivatives of the five quantities, which may be written symbolically as:

$$f = f^{(0)}[1 + (\lambda\nabla) + (\lambda\nabla)^2 + \ldots] \tag{3}$$

where λ is of the order of the mean free path and the gradient acts on the quantities n, u_i, and T. The precise form of the development is restricted by invariance considerations and can be completely determined with all the coefficients from the Boltzmann equation. Details will be omitted here, but one may perhaps point out that the systematic use of the notion of isotropic tensor field

[2] makes the calculation rather less formidable than the impression one might get from the existing literature. With the distribution function thus obtained one can calculate all the average values in (2) and one obtains *:

Zeroth approximation: Ideal hydrodynamical equations

$$P_{ij}^{(0)} = p\delta_{ij}\,, \qquad q_i^{(0)} = 0\,.$$

First approximation: Stokes-Navier equations

$$P_{ij}^{(1)} = p\delta_{ij} - 2\mu(D_{ij} - \tfrac{1}{3}D_{\alpha\alpha}\delta_{ij})\,,$$

$$q_i^{(1)} = -\nu\frac{\partial T}{\partial x_i}\,,$$

where μ and ν are given to the first approximation ‡ by the following equations:

$$\mu = \tfrac{5}{8}\sqrt{\pi mkT}\;\frac{1}{\int_0^\infty \mathrm{d}g\,g^7\,\mathrm{e}^{-g^2}\,Q_\eta(g,T)}\,,$$

$$\nu = \tfrac{25}{16}c_v\sqrt{\pi mkT}\;\frac{1}{\int_0^\infty \mathrm{d}g\,g^7\,\mathrm{e}^{-g^2}\,Q_\eta(g,T)}\,,$$

where g is the relative velocity measured in units $\sqrt{4kT/m}$, and Q_η is the so-called transport cross-section

$$Q_\eta(g,T) = 2\pi\int_0^\pi \mathrm{d}\theta\,\sin^3\theta\; I(g(4kT/m)^{\frac{1}{2}},\theta)\,.$$

The above expressions for μ and ν are exact for Maxwellian molecules, molecules repelling each other with an inverse fifth power law. The higher approximations for other molecular models can be written out but they are very complicated. For molecules

* To calculate $P_{ij}^{(n)}$ and $q_i^{(n)}$ a knowledge of $f^{(n-1)}$ is enough.

‡ The equations which determine μ and ν in terms of the molecular cross-section $I(g,\theta)$ can in general not be solved exactly, but only by a method of successive approximation which usually converges very quickly. This successive approximation method has nothing to do with the Hilberg-Enskog development (3), and should not be confused with it.

which interact with inverse sth power of the distance (repulsion), the exact expression differ from those given above only by a numerical factor which is slightly larger than unity.

Second approximation: Burnett equations

$$P_{ij}^{(2)} = p\delta_{ij} - 2\mu\,(D_{ij} - \tfrac{1}{3} D_{\alpha\alpha}\delta_{ij})$$

$$+ \omega_1 \frac{\partial u_\alpha}{\partial x_\alpha}(D_{ij} - \tfrac{1}{3} D_\alpha \delta_{ij}) + \omega_2 (L_{ij} - \tfrac{1}{3} L_{\alpha\alpha}\delta_{ij})$$

$$- \omega_2 (M_{ij} - \tfrac{1}{3} M_{\alpha\alpha}\delta_{ij}) - 2\omega_2 (N_{ij} - \tfrac{1}{3} N_{\alpha\alpha}\delta_{ij})$$

$$+ \omega_3 \left(\frac{\partial^2 T}{\partial x_i \partial x_j} - \frac{1}{3}\frac{\partial^2 T}{\partial x_\alpha \partial x_\alpha}\delta_{ij}\right)$$

$$+ \omega_4 \left(\frac{1}{2}\frac{\partial p}{\partial x_i}\frac{\partial T}{\partial x_j} + \frac{1}{2}\frac{\partial p}{\partial x_j}\frac{\partial T}{\partial x_i} - \frac{1}{3}\frac{\partial p}{\partial x_\alpha}\frac{\partial T}{\partial x_\alpha}\delta_{ij}\right)$$

$$+ \omega_5 \left(\frac{\partial T}{\partial x_i}\frac{\partial T}{\partial x_j} - \frac{1}{3}\frac{\partial T}{\partial x_\alpha}\frac{\partial T}{\partial x_\alpha}\delta_{ij}\right) + \omega_6 (Q_{ij} - \tfrac{1}{3} Q_{\alpha\alpha}\delta_{ij}) \,,$$

where

$$L_{ij} = \frac{1}{2}\frac{\partial}{\partial x_i}\left(X_j - \frac{1}{\rho}\frac{\partial p}{\partial x_j}\right) + \frac{1}{2}\frac{\partial}{\partial x_j}\left(X_i - \frac{1}{\rho}\frac{\partial p}{\partial x_i}\right),$$

$$M_{ij} = \frac{1}{2}\left(\frac{\partial u_i}{\partial x_\alpha}\frac{\partial u_\alpha}{\partial x_j} + \frac{\partial u_j}{\partial x_\alpha}\frac{\partial u_\alpha}{\partial x_i}\right),$$

$$N_{ij} = \frac{1}{2}\left[\frac{\partial u_\alpha}{\partial x_i}(D_{\alpha j} - \tfrac{1}{3} D_{\lambda\lambda}\delta_{\alpha j}) + \frac{\partial u_\alpha}{\partial x_j}(D_{\alpha i} - \tfrac{1}{3} D_{\lambda\lambda}\delta_{\alpha i})\right],$$

$$Q_{ij} = (D_{i\alpha} - \tfrac{1}{3} D_{\lambda\lambda}\delta_{i\alpha})(D_{\alpha j} - \tfrac{1}{3} D_{\mu\mu}\delta_{\alpha j}) \,,$$

and the values for the ω's are given in table 1 for Maxwellian molecules and elastic spheres.

$$q_i^{(2)} = \theta_1 \frac{\partial T}{\partial x_i}\frac{\partial u_\alpha}{\partial x_\alpha} + \theta_2 \left[\frac{2T}{3}\frac{\partial^2 u_\alpha}{\partial x_i \partial x_\alpha} + \frac{2}{3}\frac{\partial T}{\partial x_i}\frac{\partial u_\alpha}{\partial x_\alpha} + 2\frac{\partial T}{\partial x_\alpha}\frac{\partial u_\alpha}{\partial x_i}\right]$$

$$+ \theta_3 \frac{\partial p}{\partial x_\alpha}(D_{\alpha i} - \tfrac{1}{3} D_{\lambda\lambda}\delta_{\alpha i}) + \theta_4 \frac{\partial}{\partial x_\alpha}(D_{\alpha i} - \tfrac{1}{3} D_{\lambda\lambda}\delta_{\alpha i})$$

$$+ \theta_5 \frac{\partial T}{\partial x_\alpha}(D_{\alpha i} - \tfrac{1}{3} D_{\lambda\lambda}\delta_{\alpha i})$$

the values for the θ's are given in table 2.

Table 1
Values for ω

	First approximation *	Third approximation (elastic spheres)
ω_1	$\frac{4}{3}\left(\frac{7}{2} - \frac{T}{\mu}\frac{\mathrm{d}\mu}{\mathrm{d}T}\right)\frac{\mu^2}{p}$	Multiplied by 1.014
ω_2	$\frac{2\mu^2}{p}$	Multiplied by 1.014
ω_3	$\frac{3\mu^2}{\rho T}$	Multiplied by 0.806
ω_4	0	0.681
ω_5	$\frac{3\mu}{\rho T}\frac{\mathrm{d}\mu}{\mathrm{d}T} + \frac{9\mu^3}{20\rho k T^3}\sqrt{\frac{4kT}{\pi m}} \times \int_0^\infty \mathrm{d}g\, g^3\, \mathrm{e}^{-g^2}\left(g^8 - 9g^6 + \frac{63}{4}g^4\right)Q_\eta(g)$	$\frac{\mu^2}{\rho T^2}\left(0.806 \times \frac{3T}{\mu}\frac{\mathrm{d}\mu}{\mathrm{d}T} - 0.990\right)$
ω_6	$\frac{8\mu^2}{p} - \frac{16\mu^3}{35pkT}\sqrt{\frac{4kT}{\pi m}} \times \int_0^\infty \mathrm{d}g\, g^3\, \mathrm{e}^{-g^2}\left(g^6 - \frac{7}{2}g^4\right)Q_\eta(g)$	Multiplied by 0.928

* See footnote ‡ on page 7.

In tables 1 and 2, the first approximation expressions are general, true for all molecular models; the difference of the different model is absorbed in μ. For Maxwell molecules the first approximation results are exact and furthermore the integrals involving Q_η all vanish as can easily be verified since for this model $gQ_\eta(g)$ is independent of g.

The expression for $P_{ij}^{(2)}$ in its complete from was first derived by Burnett [3], and his results have been checked by several authors. The complete expression for $q_i^{(2)}$ was first given by Chapman, however, we believe that some of the θ-values of Chapman are in error [2*]. Chapman has for θ_2 a positive sign and the first

Table 2
Values for

	First approximation *	Third approximation (elastic spheres)
θ_1	$\frac{15}{4}\left(\frac{7}{2}-\frac{T}{\mu}\frac{d\mu}{dT}\right)\frac{\mu^2}{\rho T}$	Multiplied by 1.035
θ_2	$-\frac{45}{8}\frac{\mu^2}{\rho T}$	Multiplied by 1.035
θ_3	$-\frac{3\mu^2}{p\rho}$	Multiplied by 1.030
θ_4	$\frac{3\mu^2}{\rho}$	Multiplied by 0.806
θ_5	$\frac{3\mu^2}{\rho T}\left(\frac{35}{4}+\frac{T}{\mu}\frac{d\mu}{dT}\right)$ $-\frac{9}{5}\frac{\mu^3}{\rho k T^2}\sqrt{\frac{4kT}{\pi m}}\int_0^\infty dg\, e^{-g^2}(g^9-\frac{7}{2}g^7)Q_\eta$	$\frac{\mu^2}{\rho T}\left\{\frac{105}{4}\times 0.918 + 3\times 0.806\,\frac{T}{\mu}\frac{d\mu}{dT} - 0.150\right\}$

* See footnote ‡ on page 7.

part of his θ_5 is

$$\frac{3\mu^2}{\rho T}\left(\frac{5}{2}-\frac{T}{\mu}\frac{d\mu}{dT}\right) .$$

Third approximation: In this way one can go on. The authors have developed parts of the expressions for $P_{ij}^{(3)}$ and $q_i^{(3)}$, especially those terms which are linear in the derivatives of the five quantities ρ, u_i, and T.

4. DISCUSSION AND QUESTIONS

In the light of the Hilbert-Enskog theory let us look back at the general problems mentioned above:

A. To begin with we consider the least interesting problem, namely the problem of the influence of the molecular model: most calculations have been made for rather artificial molecular models (Maxwell models and elastic spheres). At present Dr. Hirschfelder in Madison is engaged in a numerical computation program to calculate the transport coefficients assuming a general Lennard-Jones type of intermolecular potential. This is now quite worthwhile, since the intermolecular potential, say for the simple inert gases is nowadays rather well known from the second virial coefficients.

The formal theory have been extended to polyatomic molecules or molecules with internal quantum states [3*]. One gets more gas coefficients, because of the possibility of exchange of energy between the translational and internal degrees of freedom. For instance the pressure tensor now gets the more general form:

$$P_{ij}^{(1)} = p\delta_{ij} - 2\mu D_{ij} - \lambda D_{\alpha\alpha}\delta_{ij}$$

or in other words the Stokes relation $2\mu + 3\lambda = 0$ is no longer fulfilled. The expressions for μ, and ν are more complicated. We find:

$$\mu = \tfrac{5}{8}\sqrt{m\pi kT}\Big(\sum_i e^{-\epsilon_i}\Big)^2$$

$$\times \Big\{\sum_{ijkl} e^{-\epsilon_i-\epsilon_j} \iint d\phi\, d\theta \sin\theta \int_0^\infty dg\, g^3 e^{-g^2}\, \Omega_{ij}^{kl}(g,\theta,\phi)$$

$$\times [g^4 \sin^2\theta - (\Delta\epsilon)^2(\sin^2\theta - \tfrac{2}{3})]\Big\}^{-1},$$

where ϵ_i is the internal energy of the molecule in the state i measured in units kT (i is written for short, it includes all the quantum numbers describing the state in question), and the summations are over all the quantum numbers, Ω_{ij}^{kl} is the collision cross-section such that two molecules in the internal quantum states i and j go after collision into states k and l, and $\Delta\epsilon$ is defined as:

$$\Delta\epsilon = \epsilon_k + \epsilon_l - \epsilon_i - \epsilon_j\,.$$

For the heat conduction coefficient we have:

$$\nu = \frac{2k^2T}{3m}\left[-\frac{15}{4}a_{10} - \frac{3}{2}\frac{\beta}{k}a_{01}\right],$$

$$a_{10} = \frac{\alpha_{10}b_{0101} - \alpha_{01}b_{0110}}{b_{1010}b_{0101} - b_{0110}^2}, \qquad a_{01} = \frac{\alpha_{01}b_{1010} - \alpha_{10}b_{0110}}{b_{1010}b_{0101} - b_{0110}^2},$$

$$\alpha_{10} = -\frac{15}{4}, \qquad \alpha_{01} = -\frac{3}{2}\frac{\beta}{k},$$

where β is the specific heat per molecule due to the internal motion. The b's are given by the following expressions:

$$b_{1010} = 4\sqrt{\frac{kT}{\pi m}}\frac{1}{(\sum_i e^{-\epsilon_i})^2}\sum_{ijkl} e^{-\epsilon_i-\epsilon_j}\iint d\phi\, d\theta \sin\theta$$

$$\times\int_0^\infty dg\, g^3 e^{-g^2}\Omega_{ij}^{kl}(g,\theta,\phi)\left[g^4\sin^2\theta + (\Delta\epsilon)^2 - \tfrac{1}{8}(\Delta\epsilon)^2\sin^2\theta\right],$$

$$b_{0110} = \frac{5}{4}\sqrt{\frac{4kT}{\pi m}}\frac{1}{(\sum e^{-\epsilon_i})^2}\sum_{ijkl} e^{-\epsilon_i-\epsilon_j}\int d\phi\, d\theta \sin\theta(\Delta\epsilon)^2$$

$$\times\int_0^\infty dg\, g^3 e^{-g^2}\Omega_{ij}^{kl},$$

$$b_{0101} = \sqrt{\frac{kT}{\pi m}}\frac{1}{(\sum_i e^{-\epsilon_i})^2}\sum_{ijkl} e^{-\epsilon_i-\epsilon_j}\int d\phi\, d\theta \sin\theta$$

$$\times\int_0^\infty dg\, g^3 e^{-g^2}\Omega_{ij}^{kl}\left[\tfrac{3}{2}(\Delta\epsilon)^2 + \{(\epsilon_i-\epsilon_j)g - (\epsilon_k-\epsilon_l)g'\}\right],$$

where g' is the relative velocity after collision in dimensionless units. Defining κ by $\kappa = \lambda - \frac{2}{3}\mu$ where $-\kappa(\partial u_\alpha/\partial x_\alpha)\delta_{ij}$ represents the contribution to the pressure tensor due to the internal motion, we have for

$$\kappa = \frac{\frac{\beta^2}{c_v^2}kT\sqrt{\frac{m\pi}{4kT}}(\sum_i e^{-\epsilon_i})^2}{\sum_{ijkl}(\Delta\epsilon)^2 e^{-\epsilon_i-\epsilon_j}\int d\phi\, d\theta\sin\theta\int_0^\infty dg\, g^3 e^{-g^2}\Omega_{ij}^{kl}(g,\theta,\phi)}.$$

This coefficient κ is closely related to the relaxation time, a time which measures the ease with which the internal motion will come to equilibrium with the translational motion. A proper definition for τ is

$$\tau = \frac{3}{2n}\frac{c_v^2}{\beta^2 kT}\kappa$$

$$= \frac{\frac{3}{2n}\sqrt{\frac{m\pi}{4kT}}(\sum_i e^{-\epsilon_i})^2}{\sum_{ijkl}(\Delta\epsilon)^2 e^{-\epsilon_i-\epsilon_j}\int d\phi\, d\theta\sin\theta\int_0^\infty dg\, g^3 e^{-g^2}\Omega_{ij}^{kl}(g,\theta,\phi)}.$$

It is not difficult to see that when there are no internal degrees of freedom or when the internal motion is not excited, the expressions for μ and ν reduce to those for monoatomic gases and κ vanishes while τ goes to infinity.

B. The question of which are the gas dynamical variables and equations seems solved. In *any* approximation the equations of motion involve *only* the five quantities ρ, u, v, w, and T; the equations are always of the first order in the time derivative, so that presumably the state of the gas is always determined if one gives at time $t = 0$ the value of ρ, u, v, w, and T as function of position. This statement constitutes the *macroscopic causality theorem* of Hilbert.

We are inclined to question this result, and would like to emphasize its paradoxical nature. Clearly in the Boltzmann equation, one needs to give at $t = 0$, f as function of $x, y, z, \xi, \eta, \zeta$ which is, of course, much more than the five moments:

$$n = \int f \, d\xi \,, \qquad u_i = \frac{1}{n} \int \xi_i f \, d\xi \,, \qquad Q = \tfrac{1}{2}\rho \, \overline{u_\alpha u_\alpha} \,.$$

Hilbert states that the macroscopic causality theorem follows only from the conditions that at any time f must be positive, finite and continuous, but his reasoning is not clear to us. It seems that the successive approximation method is so *arranged*, that the causality theorem is true, but that it is not a *proof* for it. Chapman states that perhaps any state of the gas quickly goes over into what he calls a normal state which obeys the causality theorem and that from then on the state of the gas evolves according to this theorem [4*]. But this is not proved, and of course there is also no estimate of the time required for the gas to go into a normal state.

It is clear that the theorem is certainly not valid in the limit of the Knudsen gas, because then the state of the gas depends on the number, motion and temperature of all the solid bodies inside the gas, so that the state would depend on an indefinite number of quantities. This shows perhaps also the crux of the problem.

C. What have been left out in the Hilbert-Enskog theory are the *boundary conditions*. It is clear that in the successive approximations the equations of motion become of successive higher order in the space derivatives. Because the Stokes-Navier equations are of the second order they are in harmony with the classical boundary condition of zero velocity at the solid boundary. For the Burnett equations one clearly needs more boundary conditions involving the derivatives of the velocity at the solid boundary. It is not yet clear how these boundary conditions could be obtained. One way would be to try to guess them. Consider for instance again the

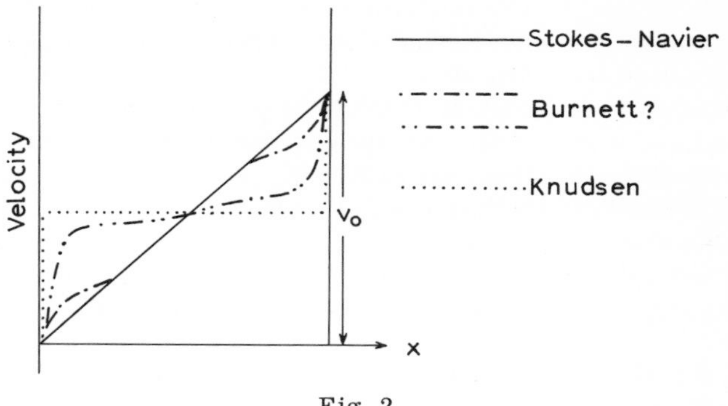

Fig. 2.

case of the Couette flow. The velocity distribution in the Stokes-Navier and Knudsen case (if there is no slip) is very simple and well known. Applying the Burnett equations to this case, one may try to choose the boundary conditions such that the solution is near Stokes-Navier and approaches Knudsen. We have tried this and a few analogous cases (flow through tubes, heat flow between plates of unequal temperature), but we have no clear-cut answer yet mainly because the equations even in these simple cases are already quite complex. It may be remarked that as seen from the figure new boundary layer phenomena are expected. This follows of course from an analogous reasoning as in the Prandtl boundary layer theory. Just as in that theory it seems not quite correct (or better perhaps not fair) to treat the Burnett terms as small perturbations. In the new boundary layer one must use the exact equations.

Of course the best way would be to derive the boundary conditions from the kinetic picture. For instance one may assume that the molecules after hitting the solid stick to the solid and are re-emitted with no memory of their past according to a Maxwell-Boltzmann distribution determined by the velocity and temperature of the wall. However, how to incorporate and harmonize such a picture with the Hilbert-Enskog development is still not clear.

5. DISPERSION OF SOUND

Clearly an experimental confirmation or disproof of the Burnett equations is lacking and would be desirable. A way which does not involve the boundary condition problem, may be obtained from the dispersion of sound and perhaps from a study of the structure of shockwaves. Primakoff [4] and later again Tsien and Schamberg [5] have pointed this out, but unfortunately their results are affected by the error in Chapman's expression for q_i which was pointed out in sect. 3. It is found that after correcting this error one gets for the velocity at angular frequency ω.

With Stokes-Navier equations:

$$V(\omega) = V_o \left[1 + \frac{141}{72} \frac{\mu^2\omega^2}{\rho^2 V_o^4}\right],$$

where ρ and V_o are the density and velocity at zero frequency. Using $\mu \sim \rho c \lambda$ it is seen that

$$\frac{\mu^2\omega^2}{\rho^2 V_o^4} \sim \left(\frac{\lambda}{\Lambda}\right)^2$$

where λ is the mean free path and Λ is the wave length of the sound wave. In fact, the expression for $V(\omega)$ given above is the first two terms in a development in powers of $(\lambda/\Lambda)^2$.

With Burnett equations

$$V(\omega) = V_o \left[1 + \frac{215}{72} \frac{\mu^2\omega^2}{\rho^2 V_o^4}\right] \tag{4}$$

for Maxwellian molecules. For elastic spheres the change in velocity is increased by 11%. The influence of the force law on the change in velocity is perhaps not very important. The dispersion effect according to Burnett is thus about 50% larger than according to Stokes-Navier treated as exact. Of course, the whole effect is small and has not been observed. For helium at 20°C one gets from equation (4)

$$\frac{V(f) - V(0)}{V(0)} \approx 2 \times 10^{-6} \left(\frac{f}{p}\right)^2,$$

with f in cycles per second and p in dynes per cm^2. For a consistent theory the coefficient of sound absorption is not changed up to this order of approximation. To obtain the next term for the coefficient of sound absorption a knowledge of $P_{ij}^{(3)}$ and $q_i^{(3)}$ is necessary.

Using the linear terms in $P_{ij}^{(3)}$ and $q_i^{(3)}$ we find

$$\alpha_{cm}^{-1} = \frac{\omega^2}{V_o^3}\left[\frac{7}{3}\frac{\mu}{\rho} - \frac{5155}{216}\frac{\mu^3\omega^2}{\rho^3 V_o^4}\right]$$

as compared to

$$\alpha_{cm}^{-1} = \frac{\omega^2}{V_o^3}\left[\frac{7}{3}\frac{\mu}{\rho} - \frac{1559}{216}\frac{\mu^3\omega^2}{\rho^3 V_o^4}\right],$$

which would follow from Stokes-Navier if treated as exact.

REFERENCES

[1] For a complete set of references see: S.Chapman and T.G. Cowling, The mathematical theory of non-uniform gases, 2nd ed. (Cambridge University Press, 1952).

[2] H.P.Robertson, Proc. Cambridge Phil. Soc. 36 (1920) 209-223.

[3] D.Burnett, Proc. London Math. Soc. 40 (1934) 382-435.

[4] H.Primakoff, Journ. of Acous. Soc. Am. 13 (1942) 14.

[5] H.S.Tsien and R.Schamberg, Journ. of Acous. Soc. Am. 18 (1946) 334.

Chapter II

ON THE DISPERSION OF SOUND IN HELIUM

C.S.WANG CHANG
University of Michigan (1948)

1. INTRODUCTION [5*]

The propagation of plane sound waves through a gas where the mean free path of the molecules is not completely negligible compared with the wave length of sound was first studied by Primakoff and later again by Tsien and Schamberg. The starting point of their calculation is the Burnett equations, i.e., the hydrodynamical equations obtained from the velocity distribution function of a non-uniform gas correct up to the second approximation according to the method of development devised by Hilbert and Enskog. In our last report [1] we pointed out how their results were in error because of an error in sign of one of the Chapman's coefficients. After correcting this it has been found that to the second order of approximation and for both the elastic spheres and the Maxwellian molecules the change of the velocity is about six or seven times larger than the results obtained from Chapman's original expressions. Furthermore the effect is about twice as large as would be obtained by treating the Stokes-Navier equations as exact. In fact the effect is of such an order of magnitude that experimental verification seems to be possible. Hence it is thought to be worthwhile to carry out the calculation for one real gas using the best known knowledge of the force law. This task has been made much easier because of the work of Dr. Hirschfelder * and his co-workers who calculated all the collision integrals $\Omega^1(n)$ necessary for the determination of the coefficients in the pressure tensor and the heat flux vector.

* We like to take this opportunity to express our gratitude to Dr. Hirschfelder for sending us their reports and permitting us to make free use of them.

The dispersion formula derived in this way by using the revised Burnett-Chapman pressure tensor and heat flux vector holds true only for a monoatomic gas where there are no internal degrees of freedom. For polyatomic gases, the additional absorption and dispersion of sound due to each internal degree of freedom are characterized by the relaxation time, τ_i, the time required for the ith internal degree of freedom to attain equilibrium with the translational motion. Only when the τ_i are known does one know whether this correction to the velocity is negligible compared to the contributions due to the viscosity μ and the heat conduction coefficient ν. From the formula for the elastic spheres and the Maxwell molecules it is seen that the latter effect will be aprreciable only when $(f/p)^2 \sim 10^5$ where f is the frequency in megacycles per second and p is the pressure in atmospheres; under atmospheric pressure this corresponds to a value of $1/f$ of the order of 10^{-9} seconds. On the other hand experiments seem to indicate that the rotational relaxation time for most gases are of the order 10^{-8} - 10^{-9} seconds. Thus in most cases the relaxation effect will mask the effects due to μ and ν. We therefore choose to study the effect in a monoatomic gas and especially helium because its interatomic force has been most extensively studied.

We will limit ourselves to classical calculations. To avoid making the necessary quantum corrections we will calculate the effect for the temperature range T = 200 to 500°K. This temperature range has the further advantage that the non-ideal behavior of helium can be neglected.

In the next section we will give the general expression for the coefficients ω and θ in the second order pressure tensor and heat flux vector, and show how they can be written formally as proportional to μ^2, $\mu\nu$, or ν^2. In sect. 3 we will give the derivation for the velocity of sound up to the second approximation. The last section will be devoted to the application to helium.

2. THE SECOND ORDER PRESSURE TENSOR AND HEAT FLUX VECTOR; THE COEFFICIENTS ω and θ

The expression for the pressure tensor has been given by both Burnett and Chapman, and in his book, Chapman also derived the heat flux vector. In our report [1] we have presented them in tensor notation. There the values of the θ's and ω's are given for elastic spheres and for Maxwell molecules. The aim of this section is mainly to give general expression for θ and ω and to show

how they are connected with the collision cross-sections. In the problem of sound propagation one usually limits oneself to small amplitudes so that only linear terms need be considered. To save writing we will restrict ourselves therefore to the linear terms. The heat flux vector and the pressure tensor can then be written as:

$$q_i^{(2)} = \theta_2 \frac{2T}{3} \frac{\partial^2 u_\alpha}{\partial x_i \partial x_\alpha} + \theta_4 \frac{\partial}{\partial x_\alpha} (D_{i\alpha} - \tfrac{1}{3} D_{\lambda\lambda} \delta_{i\alpha}) ,$$

$$p_{ij}^{(2)} = -\frac{\omega_2}{\rho} \left(\frac{\partial^2 p}{\partial x_i \partial x_j} - \frac{1}{3} \frac{\partial^2 p}{\partial x_\alpha \partial x_\alpha} \delta_{ij}\right) + \omega_3 \left(\frac{\partial^2 T}{\partial x_i \partial x_j} - \frac{1}{3} \frac{\partial^2 T}{\partial x_\alpha \partial x_\alpha} \delta_{ij}\right) . \tag{1}$$

Thus among the θ's and the ω's only θ_2, θ_4, ω_2, and ω_3 will be of interest to us. There are given by:

$$\theta_1 = -\frac{2k^2 T}{3mn^2} \int f^{(0)} A^2 c^2 \, \mathrm{d}\xi , \qquad \theta_4 = \frac{4k^2 T^2}{15mn^2} \int f^{(0)} ABc^4 \, \mathrm{d}\xi ,$$

$$\omega_2 = \frac{2kT}{15n^2} \int f^{(0)} B^2 c^4 \, \mathrm{d}\xi , \qquad \omega_3 = \frac{4k^2 T}{15mn^2} \int f^{(0)} ABc^4 \, \mathrm{d}\xi , \tag{2}$$

where A and B are functions of c, the dimensionless molecular velocity,

$$c_i = \sqrt{\frac{m}{2kT}} \, \xi_i$$

and the solutions of the integral equations:

$$nI(Ac_i) = f^{(0)}(c^2 - \tfrac{5}{2})c_i ,$$

$$nI[B(c_i c_j - \tfrac{1}{3}c^2 \delta_{ij})] = 2f^{(0)}(c_i c_j - \tfrac{1}{3}c^2 \delta_{ij}) , \tag{3}$$

I is a linear and isotropic operator defined by:

$$n^2 I(\phi) = \int \mathrm{d}\xi_1 \int \mathrm{d}\Omega \, gI(g, \theta) f^{(0)} f_1^{(0)} (\phi + \phi_1 - \phi' - \phi_1')$$

$$\equiv \int \mathrm{d}\xi_1 \int \mathrm{d}\Omega \, gI(g, \theta) f^{(0)} f_1^{(0)} \Delta\phi ,$$

where g is the relative velocity, $I(g, \theta)$ is the collision cross-section, and ϕ', ϕ_1' are the values of ϕ and ϕ_1 after collision. The integral equations (3) can not be solved in closed form. One solves them by expanding A and B in sets of orthogonal functions, the most convenient sets being the Sonine polynomials. One writes *

$$A = \sum_{r=1}^{\infty} a_r S_{\frac{3}{2}}^{(r)}(c^2)\ , \qquad B = \sum_{r=1}^{\infty} b_r S_{\frac{5}{2}}^{(r-1)}(c^2)\ . \tag{4}$$

Substituting (4) into (3), multiplying by $S_{\frac{3}{2}}^{(m)}(c^2)c_i$ and $S_{\frac{5}{2}}^{(m-1)}(c^2) \times (c_i c_j - \frac{1}{3}c^2\delta_{ij})$ respectively, contracting and integrating over $d\xi$, one obtains for both a_i and b_i an infinite set of linear equations of the form:

$$\sum_{r=1}^{\infty} a_r a_{rm} = -\frac{15}{4}\delta_{m1}\ , \qquad \sum_{r=1}^{\infty} b_r b_{rm} = 5\,\delta_{m1}\ , \tag{5}$$

where

$$a_{rm} = \int d\xi\, S_{\frac{3}{2}}^{(m)}(c^2)c_\alpha \int d\xi_1\, d\Omega\, gI(g,\theta) f^{(0)} f_1^{(0)} \Delta[S_{\frac{3}{2}}^{(r)}(c^2)c_\alpha]\ ,$$

$$b_{rm} = \int d\xi\, S_{\frac{5}{2}}^{(m-1)}(c^2)(c_\alpha c_\beta - \tfrac{1}{3}c^2\delta_{\alpha\beta}) \int d\xi_1\, d\Omega\, gI(g,\theta) f^{(0)} f_1^{(0)}$$

$$\times\, \Delta[S_{\frac{5}{2}}^{(r-1)}(c^2)(c_\alpha c_\beta - \tfrac{1}{3}c^2\delta_{\alpha\beta})]\ .$$

Both are eight fold integrals and for centrally symmetric molecules, six of these integrals can be carried out leaving only the collision integrals of the type:

$$\Omega^l(n) = \sqrt{\pi}\left(\frac{m}{4kT}\right)^{n+\frac{3}{2}} \int_0^\infty dg\, e^{-\frac{mg^2}{4kT}} g^{2n+3} \int_0^\pi d\theta \sin\theta (1 - \cos^l\theta)\, I(g,\theta)\ .$$

Chapman [2] has calculated a_{rs} and b_{rs} for r and s up to $r = s = 3$.

The linear equations (5) are solved by successive approximations. To the first approximation one stops at the first term in the sum obtaining:

$$a_1 = -\frac{15}{4a_{11}}\ , \qquad b_1 = \frac{5}{b_{11}}\ .$$

In general to the mth approximation one stops at $r = s = m$, and

* See Chapman and Cowling, The Mathematical Theory of Non-uniform Gases, p. 123 for the definitions of the Sonine polynomials $S_m^{(n)}(x)$; we will follows as much as possible the notation of Chapman and Cowling.

$$a_i = \frac{(-1)^i \frac{15}{4} \mathscr{A}_{1i}}{\mathscr{A}}, \qquad b_i = \frac{(-1)^{i-1} 5\mathscr{B}_{1i}}{\mathscr{B}},$$

where $\mathscr{A}$ is the determinant $|a_{rs}|$, and $\mathscr{A}_{1i}$ is the minor of the element a_{1i}. The a and b are functions of the temperature and they depend on the force law.

One can show that to all order of approximations the following relations always hold true:

$$a_1 = -\frac{2m}{5k^2T}\nu = -\frac{3}{2}\frac{\nu'}{kT}, \qquad b_1 = \frac{2\mu}{kT}, \tag{6}$$

where to simplify later writing we have written:

$$\nu = \frac{15}{4}\frac{k}{m}\nu',$$

where ν' has the same dimension as μ. In fact, it is exactly μ for Maxwell molecules. For other spherically symmetric molecular models

$$\nu' = s\mu .$$

s is a dimensionless temperature dependent quantity. It is different for different force laws but it is never very much different from unity. To the first approximation s is 1 for all these models. Furthermore, one can show that for molecules interacting with the inverse power law r^{-n} all the a's and b's have the same temperature dependence, so that their ratios are independent of the temperature. It is therefore convenient to write [6*]

$$\theta_2 = \epsilon_1 \frac{\nu'^2}{\rho T}, \qquad \theta_4 = T\omega_3 = \frac{\epsilon_2 \mu \nu'}{\rho}, \qquad \omega_2 = \epsilon_3 \frac{\mu^2}{p}, \tag{7}$$

where the coefficients ϵ_1, ϵ_2, and ϵ_3 are dimensionless constants independent of T for the r^{-n} law molecules. This way of writing is only formal. For other molecular models the ϵ's are still temperature dependent.

3. THE VELOCITY OF SOUND

In this section we will derive the expression for the velocity of sound from the Burnett equations. This will, of course, be mainly a repetition of known work, but to get uniformity in notation it seems not too superfluous to do it briefly again. Making use of eqs. (7) and (1), and considering only one dimensional motion one obtains for the total pressure tensor and the total heat flux vector up to second order the following expressions:

$$p_{xx} = p - \frac{4}{3}\mu \frac{\partial u}{\partial x} - \frac{2}{3}\epsilon_3 \frac{\mu^2}{p\rho}\frac{\partial^2 p}{\partial x^2} + \frac{2}{3}\epsilon_2 \frac{\mu\nu'}{\rho T}\frac{\partial^2 T}{\partial x^2},$$

$$q_x = -\frac{15k}{4m}\nu'\frac{\partial T}{\partial x} + \frac{2}{3}\epsilon_1 \frac{\nu'^2}{\rho}\frac{\partial^2 u}{\partial x^2} + \frac{2}{3}\epsilon_2 \frac{\mu\nu'}{\rho}\frac{\partial^2 u}{\partial x^2}. \tag{8}$$

The linear one dimensional Burnett equations are:

$$\frac{1}{\rho}\frac{\partial\rho}{\partial t} + \frac{\partial u}{\partial x} = 0,$$

$$\rho\frac{\partial u}{\partial t} + \frac{\partial p}{\partial x} - \frac{4\mu}{3}\frac{\partial^2 u}{\partial x^2} - \frac{2\epsilon_3}{3}\frac{\mu^2}{p\rho}\frac{\partial^3 p}{\partial x^3} + \frac{2\epsilon_2}{3}\frac{\mu\nu'}{\rho T}\frac{\partial^3 T}{\partial x^3} = 0,$$

$$\frac{3}{2T}\frac{\partial T}{\partial t} + \frac{\partial u}{\partial x} - \frac{15}{4}\frac{\nu'}{\rho T}\frac{\partial^2 T}{\partial x^2} + \frac{2\epsilon_1}{3}\frac{\nu'^2}{p\rho}\frac{\partial^3 u}{\partial x^3} + \frac{2\epsilon_2}{3}\frac{\mu\nu'}{p\rho}\frac{\partial^3 u}{\partial x^3} = 0. \tag{9}$$

Eliminating u from these equations and expression T in terms of p and ρ by the equation of state and finally introducing

$$\frac{p - p_0}{p_0} = \pi_0\, \mathrm{e}^{i(\sigma x-\omega t)}, \qquad \frac{\rho - \rho_0}{\rho_0} = r_0\, \mathrm{e}^{i(\sigma x-\omega t)}, \qquad (\pi_0, r_0 \ll 1)$$

one obtains two homogeneous equations for π_0 and r_0. Equating the determinant to zero leads then to the dispersion equation:

$$V_0^2\sigma^2 = \omega^2 + \frac{4i}{3}\frac{\mu}{\rho_0}\omega\sigma^2 + \frac{5i}{2}\frac{\nu'}{\rho_0}\omega\sigma^2 - \frac{3i}{2}\frac{\nu'}{\rho_0}V_0^2\frac{\sigma^4}{\omega}$$
$$-\frac{10}{9}\frac{\epsilon_3\mu^2}{\rho_0^2}\sigma^4 + \frac{\mu\nu'}{\rho_0^2}\sigma^4\left(\frac{8}{9}\epsilon_2 - \frac{10}{3}\right) + \frac{4\epsilon_3}{9}\frac{\nu'^2}{\rho_0^2}\sigma^4, \tag{10}$$

where V_0 is the velocity at zero frequency. Equation (10) relates the angular frequency ω to the σ, which is in general complex

$$\sigma = \sigma_1 + i\sigma_2.$$

The real part σ_1 is related to the velocity of sound while the imaginary part σ_2 gives the absorption coefficient. Equation (10) can be solved for σ by successive approximations by treating μ and ν as small quantities. This is obviously consistent with the Hilbert-Enskog development.

1) Zeroth order: Euler

$$V = V_o , \qquad \sigma_2 = 0 .$$

2) First order: Stokes-Navier

$$V = V_o , \qquad \sigma_2 = \frac{2}{3}\frac{\mu}{\rho_o}\frac{\omega^2}{V_o^3} + \frac{1}{2}\frac{\nu'}{\rho_o}\frac{\omega^2}{V_o^3} .$$

3) Second order: Burnett

$$V = V_o \left\{ 1 + \left(\frac{2}{3} + \frac{5}{9}\epsilon_3\right)\frac{\mu^2}{\rho_o^2}\frac{\omega^2}{V_o^4} + \left(\frac{5}{3} - \frac{4}{9}\epsilon_2\right)\frac{\mu\nu'}{\rho_o^2}\frac{\omega^2}{V_o^4} - \frac{\nu'^2}{\rho_o^2}\frac{\omega^2}{V_o^4}\left(\frac{3}{8} + \frac{2}{9}\epsilon_1\right)\right\},$$

$$\sigma_2 = \frac{2}{3}\frac{\mu}{\rho_o}\frac{\omega^2}{V_o^3} + \frac{1}{2}\frac{\nu'}{\rho_o}\frac{\omega^2}{V_o^3} . \tag{11}$$

Formally one can go on and on, the equation (10) can even be solved exactly. But for a consistent theory one should stop at the second approximation because we have not taken the terms of the third order in μ and ν' into account in the pressure tensor and the heat flux vector. Thus the influence of the Burnett terms on the absorption coefficient will not be seen till one includes in the equations $p^{(3)}$ and $q^{(3)}$. In the present paper we will limit ourselves to the second approximation and therefore will calculate only the velocity of sound. The second order effect due to the Stokes-Navier terms alone can be obtained from eq. (11) by putting the ϵ's equal to zero.

4. APPLICATION TO HELIUM

For the interaction between helium atoms we take the Lennard-Jones potential

$$V(r) = 4\epsilon\left[\left(\frac{r_o}{r}\right)^{12} - \left(\frac{r_o}{r}\right)^{6}\right],$$

where ϵ is the minimum for $V(r)$ and r_o is the value of the interatomic distance for which $V = 0$. Dr. Hirschfelder and his co-workers have made numerical evaluations of the collision integrals

$\Omega^l(n)$ that are necessary for the calculation of a_{rs} and b_{rs} up to $r = s = 3$ for different values of kT/ϵ.

Using de Boer and Michels' [3] values for ϵ and r_0,

$$\epsilon = 1.402 \times 10^{-15} \text{ ergs}, \qquad r_0 = 2.56 \text{ Å},$$

we have computed up to the third approximation the a's and b's for $kT/\epsilon = 20, 30, 40, 50$, corresponding to temperature 203.4, 305.1, 406.8, and 508.5^{o}K. It is found that for these temperatures and within the accuracy of the available values of $\Omega^l(n)$ the ratios a_i/a_1 and b_i/b_1 are independent of T. The result is shown in table 1. For comparison we give in table 1 also the values for Maxwell molecules and for elastic spheres. This shows that within this temperature range the helium gas behaves as simple repulsive centers of the power law $5 < n < \infty$.

Table 1

	a_2/a_1	a_3/a_1	b_2/b_1	b_3/b_1
Maxwell	0	0	0	0
Lennard-Jones	0.0672	0.0127	0.0435	0.0061
Elastic spheres	0.0944	0.0194	0.0613	0.0094

Table 2

	Maxwell	de Boer and Michels	Elastic spheres
ϵ_1	$-45/8$	$-(45/8) \times 1.008$	$-(45/8) \times 1.017$
ϵ_2	3	3×0.855	3×0.800*
ϵ_3	2	2×1.007	2×1.014
ν'/μ	1.000	1.002	1.009

* The difference between this value and the value given on p. 12 of [1] is because of the ratio ν'/μ.

Table 3

	First approx.	Second approx.	Third approx.
ϵ_1	$-45/8$	$-(45/8) \times 1.007$	$-(45/8) \times 1.008$
ϵ_2	3	3×0.862	3×0.855
ϵ_3	2	2×1.006	2×1.007

In table 2 are the values of the ϵ's and ν'/μ for the different models. To exhibit the convergence we give in table 3 the values of ϵ in the successive approximations. Putting these values for ϵ into equation (11) we find for the velocity of sound

$$V = V_0 \left[1 + \frac{230.9}{72} \frac{\mu^2 \omega^2}{\rho_0^2 V_0^4} \right],$$

or

$$\frac{\Delta V}{V_0} = \frac{230.9}{72} \frac{\mu^2 \omega^2}{\rho_0^2 V_0^4} = 44.4\ \mu^2 \left(\frac{f}{p}\right)^2, \tag{12}$$

where μ is in c.g.s. units, f in megacycles persecond and p is in atmospheres. Treating Stokes-Navier as exact would give:

$$\frac{\Delta V}{V_0} = 27.2\ \mu^2 \left(\frac{f}{p}\right)^2. \tag{13}$$

Table 4

T ($^{\circ}$K)	μ-theory (c.g.s.)	V/V_0 in % Stokes-Navier	V/V_0 in % Burnett
203.4	1.58×10^{-4}	$0.67 \times 10^{-4} (f/p)^2$	$1.10 \times 10^{-4} (f/p)^2$
305.1	2.05 "	1.14 "	1.86 "
406.8	2.47 "	1.66 "	2.71 "
508.5	2.85 "	2.21 "	3.61 "

In table 4 are the computed $\Delta V/V_0$ from equations (12) and (13). The values of μ in the second column are the theoretical values. In this temperature range they agree to within 1% of the values calculated from the empirical formula

$$\mu = 5.023\ T^{0.647} \times 10^{-6} \text{ c.g.s}$$

given by Keesom [4]*.

* A detailed discussion on the comparison of the theoretical and experimental results will be published in a paper by Dr. Hirschfelder in the Journal of Chemical Physics.

Thus we see that the effect is really not too small. For a value of $f/p = 335$ corresponding to, say, $p = 10^{-2}$ atm. and $f = 3.35$ mc and $T = 305.1^{\circ}$K

$$\frac{\Delta V}{V_0} = 20.8\%$$

in contrast to the 3% obtained by Tsien and Schamberg. Of course, for $\Delta V/V_0$ so large the next term in the development will not be negligible. To get the next term in the velocity one would have to calculate $p^{(4)}$ and $q^{(4)}$.

REFERENCES

[1] Chap. I of this book.
[2] Chapman and Cowling, The Mathematical Theory of Non-uniform Gases, p. 161.
[3] J.de Boer and Michels, Physica 5 (1938) 945; 6 (1939) 409.
[4] W.H.Keesom, Helium, p. 107.

Chapter III

ON THE THEORY OF THE THICKNESS OF WEAK SHOCK WAVES

C.S.WANG CHANG
University of Michigan (1948)

1. INTRODUCTION [6 *]

The two problems often suggested in connection with the Burnet equations are: (1) the velocity of propagation of high-frequency sound waves, and (2) the shock-wave problem. In the Burnett equations higher-order differential quotients appear; thus the application of Burnett equation to most problems calls for a knowledge of additional boundary conditions. The above-mentioned two problems are the simplest because they are the two problems that do not involve boundary conditions. The first problem has been treated by Primakoff and Tsien and Schamberg, and recently we have considered especially the calculation of the velocity of sound in helium gas [1]. The problem of the thickness of one-dimensional shock waves has been studied by both Becker [2] and Thomas [3] using Stokes-Navier equations as exact, though Thomas included in his paper a discussion as to the changes to be expected when the Burnett terms are taken into account. Becker started with the Stokes-Navier equations and treated the viscosity coefficient, μ, and the heat conduction coefficient, ν, as constants independent of the temperature. For a particular value of the ratio $f = \nu/c_v\mu$, namely $f = 4\gamma/3$ where γ is the ratio of the specific heats c_p and c_v of the gas, he was able to solve the hydrodynamical equations exactly and he obtained an expression for the velocity distribution. From this velocity distribution the shock thickness was calculated. His results show that for the Mach number, M, slightly larger than unity, the shock-wave thickness, t, is about a few mean free paths, while for large Mach numbers, say $M \sim 35$, t is of the order of 10^{-7} cm, or of the order of the mean distance between the molecules.

There are three objections to this theory:

(1) The Stokes-Navier equations were treated as exact, which is against the spirit of the development theory of Enskog.

(2) The viscosity and the heat conduction coefficients are certainly not constant especially for very strong shocks where the temperature change is tremendous.

(3) His solution is for the one value of $f = 4\gamma/3$. For diatomic gases with $\gamma = 7/5$, this value of f (1.87) is near enough to both the theoretical and the experimental value, so that the calculation seems to be applicable for air. But the Stokes-Navier equations are strictly true only for monoatomic gases where there are no internal degrees of freedom. For polyatomic gases the relaxation time is usually so large that its effect masks the effect of the other two gas coefficients. Thus we do not expect results derived from the Stokes-Navier equations to hold good for air, or any polyatomic gas unless we are sure that the relaxation time is very short.

Thomas' calculation removes the second objection. He included in his calculation the temperature dependence of μ and ν. An elastic sphere model was taken for the molecules, so that both μ and ν were taken to be proportional to the square root of the absolute temperature. The equations are still soluble for the same particular value of f. It was found that for $M=2$ the thickness of the shock wave is about four times the mean free path, λ_m (the mean free path at the place where the velocity gradient is a maximum). For the infinity strong shock, t has the value of $1.74\,\lambda_m$. We will discuss his results and calculation further in sect. 2.

Recently Mott-Smith * made an essentially different approach to calculate the thickness of shock waves. Using neither the Stokes-Navier equations nor the idea of Enskog's development, he started directly from the Boltzmann equation and the Boltzmann's equation of transport. For both monoatomic and diatomic gases, he found that the thickness of a shock waves is inversely proportional to $(M-1)$ for $(M-1)$ small, and that for infinitely strong shock waves, t is of the order of a few mean free paths.

As a summary of the existing results on the calculation of the shock thickness, we have plotted in fig. 1 the values of λ/t against the Mach number M, where λ is the mean free path in the medium before the shock. Curve I is computed from Becker's theory,

* We are indebted to Commander H. M. Mott-Smith for showing us his manuscript on the kinetic theory treatment of a shock wave. Later published in the Physical Review, Vol. 82 (1951) p. 885.

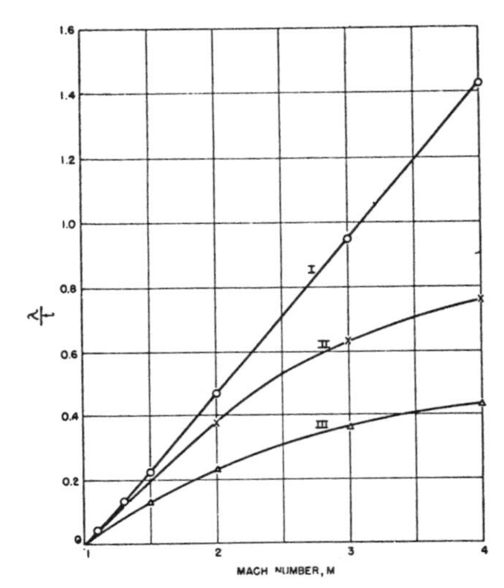

Fig. 1. Variation of the thickness of shock waves with the Mach number.

curve II follows from the calculation of Thomas except that t is now measured in terms of λ instead of λ_m. Curve III is calculated by using the formulas of Mott-Smith *. Curves I and II have the same initial slope. For high M according to Becker λ/t increases almost linearly with M, whereas according to Thomas λ/t increases much shower with M and approaches a value of $\lambda/t = 1.02$ for $M = \infty$. Curve III has a smaller initial slope and the asymptotic value of λ/t is 0.55. One sees that there is quite a difference between the theories especially for strong shock waves. In our opinion, for strong shock waves the calculation of Mott-Smith is probably the most dependable, but his theory has to be modified since an exact theory should give the initial slope of Becker and Thomas.

We propose to develop a consistent theory for the shock thickness for weak shock waves. We shall follow the idea of the Enskog development and develop the shock thickness in a power series of $M - 1$. To the first approximation we shall use the Stokes-Navier equations and stop at the first power of $(M - 1)$. We shall see that

* We differ here from Mott-Smith in one respect. We found that in conformity with Becker's definition of the shock thickness $t = 4\lambda/B$ instead of Mott-Smith's $t = \lambda/B$. For B we have used Mott-Smith's value determined from the consideration of the transport of u^2, where u is the streaming velocity, and $\gamma = 7/5$.

we get the same results as Thomas if we develop Thomas' final result also in a power series in $(M-1)$. To the second order we include the Burnett terms and carry the calculation to $(M-1)^2$, etc.

In sect. 2 we shall restrict ourselves to the first-order calculation. We will give Thomas' result and compare it with ours. The calculation will be extended to the second- and thord-order terms in sect. 3. The last section will be devoted to the application of the result to a monoatomic gas.

2. THE THICKNESS OF SHOCK WAVES FROM THE STOKES-NAVIER EQUATIONS

We start by writing down the general one-dimensional hydrodynamical equations for the steady state,

$$\frac{\mathrm{d}}{\mathrm{d}x}(\rho u) = 0 \ ,$$

$$\frac{\mathrm{d}p}{\mathrm{d}x} + \rho u \frac{\mathrm{d}u}{\mathrm{d}x} = 0 \ ,$$

$$u\frac{\mathrm{d}E}{\mathrm{d}x} + E\frac{\mathrm{d}u}{\mathrm{d}x} + \frac{\mathrm{d}q}{\mathrm{d}x} + p\frac{\mathrm{d}u}{\mathrm{d}x} = 0 \ ,$$

where ρ is the density, u is the streaming velocity, p is the pressure tensor ($p = p_{xx}$ in the present case), and q is the heat current density. E is the total energy $= nkT/(\gamma - 1)$. These equations can be integrated once immediately, yielding

$$\rho u = \text{const.} = A \ ,$$

$$p + Au = \text{const.} = B \ ,$$

$$uE + q + Bu - \tfrac{1}{2}Au^2 = \text{const.} = C \ , \tag{1}$$

where A, B, and C are constants. Given the velocity and the state of the gas before the shock, equations (1) determine the conditions after the shock uniquely. In the zeroth order (ideal hydrodynamical equations) one must put $p = nkT$, $q = 0$, and equations (1) then give the well-known shock conditions. The shock wave appears as a discontuinity in the velocity, the temperature, the density, and the pressure. However, when the first-order pressure tensor and heat flux vector are taken into account one finds that the shock waves are really not discontinuities but there are regions in which the velocity, the density, the temperature, and the pressure undergo large changes. In this order of approximation (Stokes-

Navier equations),

$$p = nkT - \frac{4}{3}\mu\frac{\mathrm{d}u}{\mathrm{d}x},$$

$$q = -\nu\frac{\mathrm{d}T}{\mathrm{d}x},$$

equations (1) become now:

$$\rho u = A\,, \tag{2a}$$

$$\rho\frac{kT}{m} - \frac{4}{3}\mu\frac{\mathrm{d}u}{\mathrm{d}x} + \rho u^2 = B\,, \tag{2b}$$

$$\frac{\gamma}{\gamma-1}\rho u\frac{kT}{m} - \frac{4}{3}\mu u\frac{\mathrm{d}u}{\mathrm{d}x} + \tfrac{1}{2}\rho u^3 - \nu\frac{\mathrm{d}T}{\mathrm{d}x} = C\,. \tag{2c}$$

The constants A, B, and C can be expressed in terms of the gas quantities before the shock (indicated by the subscript zero):

$$A = \rho_0 u_0\,,$$

$$B = \rho_0\frac{kT_0}{m} + \rho_0 u_0^2\,,$$

$$C = \frac{\gamma}{\gamma-1}\rho_0 u_0\frac{kT_0}{m} + \tfrac{1}{2}\rho_0 u_0^3\,.$$

Equations (2) will be expressed in dimensionless forms. For this purpose we introduce the following dimensionless variables,

$$\tau = \frac{T}{T_0}\,, \qquad v = \frac{u}{u_0}\,, \qquad b = \frac{x}{t}\,.$$

We will also need the Mach number,

$$M = \frac{u_0}{a_0}\,,$$

where a_0 is the velocity of sound in the medium before the shock. For the temperature dependence of the viscosity coefficient we will first use the more general expression

$$\frac{\mu}{\mu_0} = \left(\frac{T}{T_0}\right)^{1/s},$$

where s is equal to one or two for Maxwell molecules (points repelling with a force $\sim 1/r^5$), and elastic spheres, respectively. The specification of the value of s will be made later when we apply our results to definite molecular models. Further we write

$$\mu_0 = \eta \rho_0 a_0 \lambda \, .$$

The expression η is the dimensionless constant $5\sqrt{2}\pi/16\sqrt{\gamma}$, and λ is a general Maxwell mean free path in the medium before the shock, defined by

$$\lambda = \frac{\sqrt{2}}{n_0 \int_0^\infty g^7 e^{-g^2} Q_\eta(g)\, dg} ,$$

n being the number density and $Q_\eta(g)$ is the so-called transport cross section. For elastic spheres

$$Q_\eta = \frac{2\pi\sigma^2}{3} ,$$

where σ is the molecular diameter; λ is simply the ordinary Maxwell mean free path. Dividing the equations (2b) and (2c) by (2a) expressing the resulting equations in dimensionless variables we obtain:

$$\tau - \frac{4}{3}\eta\gamma M \frac{\lambda}{t} \tau^{1/s} v \frac{dv}{db} + \gamma M^2 v(v-1) - v = 0 \tag{3a}$$

and

$$\tau - \frac{4}{3}\eta(\gamma-1) M \frac{\lambda}{t} \tau^{1/s} v \frac{dv}{db} - \frac{\eta f}{\gamma} M^{-1} \frac{\lambda}{t} \tau^{1/s} \frac{d\tau}{db} + \tfrac{1}{2}(\gamma-1) M^2 v^2 - \tfrac{1}{2}(\gamma-1) M^2 - 1 = 0 \, . \tag{3b}$$

For further calculations it is simpler to use equation (3a) and the equation obtained by subtracting (3b) from (3a). These are

$$\tau - \frac{4}{3}\eta\gamma M \frac{\lambda}{t} \tau^{1/s} v \frac{dv}{db} + \gamma M^2 v(v-1) - v = 0 \, , \tag{4a}$$

$$\frac{4}{3}\eta M \frac{\lambda}{t} \tau^{1/s} v \frac{dv}{db} - \frac{\eta f}{\gamma} M^{-1} \frac{\lambda}{t} \tau^{1/s} \frac{d\tau}{db} + \tfrac{1}{2}(\gamma-1) M^2 v^2 - \tfrac{1}{2}(\gamma-1) M^2 - \gamma M^2 v(v-1) + (v-1) = 0 \, . \tag{4b}$$

Making now series expansions of all the quantities in power series in y where y stands for $M - 1$,

$$\tau = 1 + \tau_1 y + \tau_2 y^2 + \ldots ,$$

$$v = 1 + v_1 y + v_2 y^2 + \ldots ,$$

$$\frac{\lambda}{t} = g_1 y + g_2 y^2 + \ldots . \qquad (5)$$

The expansions for τ and v start from the constant term 1 while that for λ/t starts with the linear term $g_1 y$ because for y equals zero there is no shock wave. Substituting eq. (5) into (4a) and equating terms of equal powers of y one finds that the terms independent of y are identically zero while the first-order terms yield:

$$\tau_1 = -(\gamma - 1) v_1 .$$

From eq. (4b) and the above equation one finds the differential equation for v_1

$$\eta g_1 \left(\frac{4}{3} + f - \frac{f}{\gamma}\right) \frac{dv_1}{db} = 2v_1 \left(1 + \frac{\gamma+1}{4} v_1\right) ,$$

the solution of which is

$$v_1 = \frac{A\, e^{2\alpha b}}{1 - \frac{\gamma+1}{4} A\, e^{2\alpha b}} , \qquad (6)$$

where

$$\alpha = \frac{1}{\eta g_1 \left(\frac{4}{3} + f - \frac{f}{\gamma}\right)} .$$

For $b = -\infty$, $v_1 = 0$, as it should since one is then in the region before the shock. The value of v_1 at $b = +\infty$ is given by the term $\sim y$ of the expansion of the velocity after the shock, v_f, determined by the shock conditions,

$$v_f = \frac{1}{\gamma+1} \left(\gamma - 1 + \frac{2}{M^2}\right) = 1 - \frac{2}{\gamma+1} (2y - 3y^2 \ldots) , \qquad (7)$$

i.e., $v_1 = -4/(\gamma + 1)$ at $b = +\infty$. Equation (6) satisfies therefore both the boundary conditions. The integration constant A remains undetermined because the shock is not localized. If desired, one can fix A by requiring that the maximum velocity gradient occus at b = 0. This is, however, arbitrary and it is not necessary at present.

Following the earlier workers we define the shock-wave thickness by

$$t = (u_f - u_o) \Big/ \left(\frac{du}{dx}\right)_{max} . \tag{8}$$

In dimensionless notation this becomes:

$$\left(\frac{dv}{db}\right)_{max} = (v)_f - 1 ,$$

where the right-hand side is known and the left-hand side involves the parameters g_1, g_2, etc., which measures the thickness. To the present order of approximation we have therefore

$$\left(\frac{dv_1}{db}\right)_{max} = -\frac{4}{\gamma+1} .$$

From the velocity distribution (6) it follows that the maximum slope occurs at the value of b satisfying the following equation:

$$e^{2\alpha b} = -\frac{4}{(\gamma+1)A} .$$

Using this relation and equations (6) and (8), one can solve for α and g_1:

$$\alpha = 2$$

$$g_1 = \frac{8\sqrt{\gamma}}{5\sqrt{2\pi}} \frac{1}{\frac{4}{3} + f - \frac{f}{\gamma}} . \tag{9}$$

We observe that g_1 is independent of s, the temperature dependence of μ and ν. This is not surprising since it is clear that the temperature dependence of μ and ν will contribute only to the second-order terms. Up to this order the velocity distribution is symmetric and the velocity at the maximum velocity gradient is $-2y/(\gamma+1)$.

By eliminating x from eqs. (2a) and (2c) Becker arrived at a differential equation connecting T and u which can be solved for $f = 4\gamma/3 = 1.87$. This relation of T as a function of u was put back into eq. (2b) and and expression for du/dx was obtained which depends only on the unknown u. Making use of this last expression the calculation of the shock thickness is straightforward. Becker did not take into account the temperature variation of the viscosity and the heat conduction coefficients, but the essential features of the calculation are not changed by the inclusion of these temperature dependences. Letting μ and ν be proportional to $T^{1/s}$, the generalized Becker's relation between du/dx and u is, in our notation and with λ = the mean free path before the shock:

$$\frac{4}{3}\eta\lambda\frac{dv}{dx}+\frac{1}{\gamma Mv}\frac{v^2+v_f-v(1+v_f)}{\left(\frac{1+v_f}{1+\gamma M^2}\right)^{1-1/s}\left\{v_f-\left[\frac{\gamma M^2}{1+\gamma M^2}(1+v_f)-1\right]v^2\right\}^{1/s}} .$$

This reduces to Becker's result if $s = 0$, while for $s = 2$ it is the Thomas' formula. Restricting ourselves to $s = 2$, it can be deduced that dv/dx will be a maximum when v is a solution of the cubic equation:

$$v^3(v_f+1)\left[\frac{\gamma M^2}{1+\gamma M^2}(1+v_f)-1\right]-v^2v_f\left[2\frac{\gamma M^2}{1+\gamma M^2}(1+v_f)-1\right]+v_f^2=0 .$$

Thomas computed the value of λ/t for several values of the Mach number using eq. (8) as the defining equation for t. His results are exact when the Stokes-Navier equations are considered to be exact. To compare with our results we make again series expansions in powers of y. We find

$$\frac{\lambda}{t}=\frac{6}{5\sqrt{2\pi\gamma}}\,y\left(1+\frac{5-3\gamma}{2(1+\gamma)}\,y+\frac{9\gamma^2-34\gamma+21}{4(\gamma+1)^2}\,y^2+\ldots\right) .$$

One sees that, as to be expected, the coefficient of y is the same as g_1 if in equation (9) we put $f = 4\gamma/3$.

3. HIGHER APPROXIMATIONS

To extend this calculation so that it is applicable to stronger shock waves, it is necessary to make the development to higher-order terms. However, it is not correct merely to solve for v_2, v_3, etc., from equations (4). A theory consistent with Enskog's expansion idea must then include in the pressure tensor and the heat conduction vector the higher-order terms resulting from the Enskog development. In this section the calculation will be extended two steps further.

3.1. *The second approximation* (*the Burnett terms*)

It is clear from the method of development that when one goes to the next approximation one needs only to take the linear second-order Burnett terms into account. The nonlinear terms contribute only to the third- or higher-order of approximation just as the temperature dependence of μ and ν does not contribute to the first-order effect. The linear terms of $p_{xx}^{(2)}$ and $q_x^{(2)}$ are:

$$p_{xx}^{(2)} = \frac{2}{3}\,\omega_2\,\frac{\mu^2}{\rho u}\,\frac{\mathrm{d}^2 u}{\mathrm{d}x^2} + \frac{2}{3}\,\frac{\mu^2}{\rho T}\,(\omega_3 - \omega_2)\,\frac{\mathrm{d}^2 T}{\mathrm{d}x^2}$$

$$q_x^{(2)} = \frac{2}{3}\,(\theta_2 + \theta_4)\,\frac{\mu^2}{\rho}\,\frac{\mathrm{d}^2 u}{\mathrm{d}x^2}\,,$$

where the ω's and the θ's are slowly varying functions [4] of T. They are dimensionless quantities, and for the Maxwellian model they are pure numbers independent of T. The two equations taking the place of eq. (4) are:

$$\tau - \frac{4}{3}\gamma\eta M\,\frac{\lambda}{t}\,v\tau^{1/s}\,\frac{\mathrm{d}v}{\mathrm{d}b} + \frac{2}{3}\,\omega_2\gamma\eta^2\,\frac{\lambda^2}{t^2}\,v\tau^{2/s}\,\frac{\mathrm{d}^2 v}{\mathrm{d}b^2}$$

$$+ \frac{2}{3}\,(\omega_3 - \omega_2)\gamma\eta^2\,\frac{\lambda^2}{t^2}\,v^2\tau^{(2-s)/s}\,\frac{\mathrm{d}^2\tau}{\mathrm{d}b^2} + \gamma M^2 v(v-1) - v = 0\,. \qquad (10a)$$

and

$$\frac{4}{3}\,\eta\,\frac{\lambda}{t}\,v\tau^{1/s}\,\frac{\mathrm{d}v}{\mathrm{d}b} - \frac{f\eta}{\gamma}\,\frac{1}{M}\,\frac{\lambda}{t}\,\tau^{1/s}\,\frac{\mathrm{d}\tau}{\mathrm{d}b} - \frac{2}{3}\,\omega_2\eta^2\,\frac{\lambda^2}{t^2}\,v\tau^{2/s}\,\frac{\mathrm{d}^2 v}{\mathrm{d}b^2}$$

$$+ \frac{2}{3}\,(\theta_2 + \theta_4)\,\eta^2(\gamma-1)\,\frac{\lambda^2}{t^2}\,v\tau^{2/s}\,\frac{\mathrm{d}^2 v}{\mathrm{d}b^2} - \frac{2}{3}\,(\omega_3 - \omega_2)\,\eta^2\,\frac{\lambda^2}{t^2}\,v^2\tau^{(2-s)/s}\,\frac{\mathrm{d}^2\tau}{\mathrm{d}b^2}$$

$$+ \tfrac{1}{2}v^2(\gamma-1)M^2 - \tfrac{1}{2}(\gamma-1)M^2 - \gamma M^2 v(v-1) + v - 1 = 0\,. \qquad (10b)$$

From eq. (10a) it follows that

$$\tau_2 = -(\gamma-1)v_2 - \gamma v_1(v_1 + 2) + \tfrac{4}{3}\,\eta\gamma g_1 v_1'\,.$$

This equation together with eq. (10b) gives the differential equation for v_2 in terms of known functions,

$$v_2' - 2\alpha\left(1 + \frac{\gamma+1}{2}\,v_1\right)v_2 = \alpha v_1 + \alpha(\gamma+1)v_1^2$$

$$- \left(\frac{g_2}{g_1} - \frac{\frac{4}{3} + f + \frac{f}{\gamma}}{\frac{4}{3} + f - \frac{f}{\gamma}}\right)v_1' + \left(\frac{\gamma-1}{s} - \frac{2f + \frac{4}{3}}{\frac{4}{3} + f - \frac{f}{\gamma}}\right)v_1 v_1' + \frac{\frac{4}{3}f + \epsilon}{\alpha\left(\frac{4}{3} + f - \frac{f}{\gamma}\right)^2}\,v_1''\,, \qquad (11)$$

where

$$\epsilon = \tfrac{2}{3}\omega_2 - \tfrac{2}{3}(\omega_3 - \omega_2)(\gamma - 1) - \tfrac{2}{3}(\theta_2 + \theta_4)(\gamma - 1) .$$

If one is only interested in the thickness of the shock waves, it will not be necessary to solve the differential equation (11). Since v_1 is given by eq. (6), by inspection it is seen that the integrating factor of eq. (11) is

$$e^{-2\alpha b}\left(1 - \frac{\gamma+1}{4} A\, e^{2\alpha b}\right)^2 .$$

Thus

$$v_2 = \frac{e^{2\alpha b}}{\left(1 - \frac{\gamma+1}{4} A\, e^{2\alpha b}\right)^2} \int e^{-2\alpha b}\left(1 - \frac{\gamma+1}{4} A\, e^{2\alpha b}\right)^2 F\, db$$

with F standing for the right-hand members of eq. (11). But

$$\frac{d}{db} \frac{e^{2\alpha b}}{\left(1 - \frac{\gamma+1}{4} A\, e^{2\alpha b}\right)^2} = 0$$

at the maximum slope; hence we have

$$(v_2')_{max} = (F)_{max} ,$$

where the subscript max means that all quantities are to be evaluated at b_m where v_2'' is zero. The g_2 is then solved by setting

$$(v_2')_{max} = +\frac{6}{\gamma+1} ,$$

as follows from equations (7) and (8).

The newly introduced Burnett terms enter in F only as coefficients of v_1'', but $(v_1'')_{max}$ vanishes; hence the Burnett terms do not contribute to g_2, or the shock-wave thickness to this order of approximation. The ratio g_2/g_1 is found to be:

$$\frac{g_2}{g_1} = \frac{3-\gamma}{2(\gamma+1)} - \frac{2(\gamma-1)}{s(\gamma+1)} . \tag{12}$$

For $s = 2$, eq. (12) agrees with the coefficient of the y^2 term of the result of Thomas. However, this fact should not be taken too seriously, since it is due to the definition (8) of the shock-wave thickness t. If another definition for the thickness was chosen, the Burnett terms would have influenced the coefficient of y^2 since they have an influence on the velocity distribution inside the shock. The velocity distribution is obtained by integrating equation (11) and one finds:

$$v_2 = \frac{e^{2\alpha b}}{\left(1 - \frac{\gamma+1}{4} A\, e^{2\alpha b}\right)^2} \Bigg\{ \alpha A b - \frac{1}{2}\frac{\gamma+1}{4} A^2 e^{2\alpha b} + \frac{A^2}{2} (\gamma+1)\, e^{2\alpha b}$$

$$\frac{\frac{4}{3}\left(\frac{g_2}{g_1} - 1\right) + \frac{8}{3} + \frac{f}{\gamma}(\gamma-1)\left(\frac{g_2}{g_1} - 1\right) + 2f}{\frac{4}{3} + f - \frac{f}{\gamma}} 2\alpha A b$$

$$- \left(\frac{\gamma-1}{s} - \frac{2f + \frac{4}{3}}{\frac{4}{3} + f - \frac{f}{\gamma}} \right) \frac{4A}{\gamma+1} \log\left(1 - \frac{\gamma+1}{4} A\, e^{2\alpha b}\right)$$

$$+ \frac{\frac{4}{3} f + \epsilon}{\alpha\left(\frac{4}{3} + f - \frac{f}{\gamma}\right)^2} \left[4\alpha^2 A b - 4\alpha A \log\left(1 - \frac{\gamma+1}{4} A\, e^{2\alpha b}\right)\right] + C \Bigg\}, \tag{13}$$

where the integration constant is to be determined from the fact that at b_m, v_2'' must vanish. In the last section we will plot the velocity distributions $v_1 y + v_2 y^2$ both from the consistent theory and from the Stokes-Navier equations when treated as exact.

3.2. *Third approximation*

In this approximation not only should all the Burnett terms be taken into account but also the linear terms arising from $f^{(3)}$ are of importance. Dropping all the terms that will not contribute, we write

$$p_{xx}^{(2)} = \frac{2}{3}\omega_2 \frac{\mu^2}{\rho u}\frac{d^2 u}{dx^2} + \frac{2}{3}(\omega_3 - \omega_2)\frac{\mu^2}{\rho T}\frac{d^2 T}{dx^2} + \frac{2}{3}(\omega_2 - \omega_4)\frac{\mu^2}{\rho u T}\frac{dT}{dx}\frac{du}{dx}$$

$$+ \frac{2}{3}(\omega_4 + \omega_5)\frac{\mu^2}{\rho T^2}\left(\frac{dT}{dx}\right)^2 + \frac{2}{3}\left(\omega_1 - \tfrac{7}{3}\omega_2 + \tfrac{4}{9}\omega_6\right)\frac{\mu^2}{\rho}\left(\frac{du}{dx}\right)^2$$

$$- \frac{2}{3}\omega_2 \frac{\mu^2}{\rho u^2}\left(\frac{du}{dx}\right)^2,$$

$$q_x^{(2)} = \frac{2}{3}(\theta_2+\theta_4)\frac{\mu^2}{\rho}\frac{d^2u}{dx^2} + (\theta_1+\tfrac{8}{3}\theta_2+\tfrac{2}{3}\theta_3+\tfrac{2}{3}\theta_5)\frac{\mu^2}{\rho T}\frac{dT}{dx}\frac{du}{dx}$$

$$-\frac{2}{3}\theta_3\frac{\mu^2}{\rho u}\left(\frac{du}{dx}\right)^2 ,$$

$$p_{xx}^{(3)} = \frac{2}{3}(\omega_7+\omega_8)\frac{\mu^3}{p\rho}\frac{d^3u}{dx^3} ,$$

$$q_x^{(3)} = (\theta_6+\theta_7)\frac{\mu^3}{\rho^2 T}\frac{d^3T}{dx^3} - \theta_6\frac{\mu^3}{\rho^2 u}\frac{d^3u}{dx^3} ,$$

where ω_7, ω_8, θ_6, and θ_7, like the other ω's and θ's are the new slowly varying temperature-dependent, dimensionless quantities, For Maxwellian molecules they are simply numbers. The calculation is tedious and lengthy but the method is the same. We shall only present the final result in the most general form:

$$\frac{g_3}{g_1} = \frac{\gamma^2-6\gamma+1}{2(\gamma+1)^2} - \frac{\frac{8}{3}f}{(\gamma+1)B^2} + \frac{(2f+\frac{4}{3})^2}{(\gamma+1)^2B^2} - \frac{(\frac{4}{3}f+\epsilon)^2}{B^4}$$

$$+\frac{2(\gamma^2-6\gamma+1)}{s(\gamma+1)^2} + \frac{\frac{8}{3}f(\gamma-1)}{s(\gamma+1)B^2} + \frac{8}{3sB} + \frac{3(\gamma-1)^2}{s^2(\gamma+1)^2}$$

$$-\frac{4}{3(\gamma+1)B^2}\Big[\gamma\omega_1 - \tfrac{4}{3}\gamma\omega_2 - \frac{(f+\frac{4}{3})\gamma(\gamma+1)}{B}\omega_2 + \gamma(\gamma-1)\omega_3 + \gamma(\gamma-1)\omega_4$$

$$+(\gamma-1)^2\omega_5 + \tfrac{4}{9}\gamma\omega_6 + \tfrac{3}{2}(\gamma-1)^2\theta_1 + 4(\gamma-1)^2\theta_2 + \gamma(\gamma-1)\theta_3 + (\gamma-1)^2\theta_5\Big]$$

$$-\frac{2}{B^3}\left[\tfrac{2}{3}\gamma(\omega_7+\omega_8) + \gamma(\gamma-1)\theta_6 + (\gamma-1)^2\theta_7\right] ,$$

in which we have put B for

$$\frac{4}{3} + f - \frac{f}{\gamma} .$$

4. APPLICATION TO MONOATOMIC GASES

The calculation will now be applied to monoatomic gases. We shall limit ourselves to the treatment of the Maxwellian model, the molecules interacting with the r^{-5} force law. This choise was made not becayse this model is any better than other models from the physical point of view but because using this model the task of numerical computation will be much simplified. For Maxwell molecules the constants have the following values:

$$f = \tfrac{5}{2}, \quad \gamma = \tfrac{5}{3}, \quad B = \tfrac{4}{3} + f - \frac{f}{\gamma} = \tfrac{7}{3},$$

$$\omega_1 = \tfrac{10}{3}, \quad \theta_1 = \tfrac{75}{8},$$

$$\omega_2 = 2, \quad \theta_2 = -\tfrac{45}{8}, \quad \omega_7 = -\tfrac{4}{3},$$

$$\omega_3 = 3, \quad \theta_3 = -3, \quad \omega_8 = \tfrac{5}{3},$$

$$\omega_4 = 0, \quad \theta_4 = 3, \quad \theta_6 = -\tfrac{5}{8},$$

$$\omega_5 = 3, \quad \theta_5 = \tfrac{117}{4}, \quad \theta_7 = \tfrac{21}{16}.$$

$$\omega_6 = 8,$$

The result for the shock thickness is summarized in table 1.

Table 1
Shock thickness in monoatomic gases (Maxwell molecules)

	g_2/g_1	g_3/g_1
Stokes-Navier	-1/4	-0.349
Burnett	-1/4	-1.176
Third-order equations	-1/4	-1.271

$$\frac{\lambda}{t} = g_1 y \left(1 + \frac{g_2}{g_1} y + \frac{g_3}{g_1} y^2 + \dots\right), \qquad g_1 = \frac{8}{7\sqrt{2\pi\gamma}},$$

λ = the mean free path in the medium before the shock,

$y = M - 1$, $\qquad M$ = Mach number .

From this table one sees that the development of the thickness of shock waves in powers of $(M-1)$ converges very slowly and is therefore only applicable for Mach numbers which are only slightly bigger than one *. It should also be remembered that this calculation is valid only for monoatomic gases. The extension to diatomic gases is possible. If the effect of the relaxation time is of the same order of magnitude as the effects of μ and ν, one can take this fact into account to the first order of approximation by

* See footnote on next page.

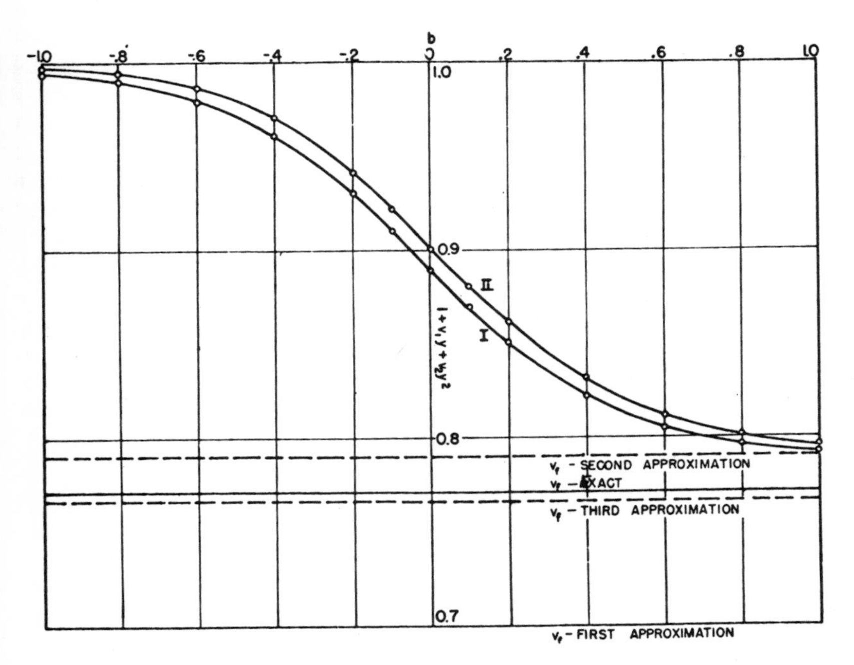

replacing $4/3\ \mu$ in the expression for $p_{xx}^{(1)}$ by $4/3\ \mu + \kappa$ where κ is the "second viscosity coefficient" ‡. To higher orders not only the expressions for ω's and θ's will be changed but also new constants

* In fact one sees already from the series expansion of the velocity after the shock, v_f, that one probably would not get fast convergent series for λ/t. The expression v_f is given by

$$v_f = 1 - \frac{2}{\gamma+1}(2y - 3y^2 \ldots) ,$$

which is convergent only for $y < 1$ and even then the convergence is very slow. Since the consistent theory is certainly to yield the successive terms in the above formula for v_f in the successive approximations, it is seen that so far as finding the velocity distribution is concerned we cannot expect to find better convergence. The shock thickness is derived from the velocity distribution, and hence the slow convergence of our result could have been anticipated.

‡ See ref. [4].

will enter. Because of the slow convergence of the development it seems at present hardly worthwhile to go into such calculations.

In sect. 3 it was remarked that the fact that g_2/g_1 is not changed by taking into account of the Burnett terms is only accidental. There are certainly second-order effects due to the Burnett terms, only these effects are not reflected in our calculation of the shock-wave thickness because of the particular definition we adopted. To see one of the second-order effects of the Burnett terms we have plotted in fig. 2 the curves with $v_1 y + v_2 y^2$ against b. Curve I is calculated from the Stokes-Navier equations while curve II is obtained from the Burnett equations. The horizontal dotted lines are the asymptotes of the values of $v_f = u_f/u_o$ up to the different orders of approximations as indicated. The plot is made for $y = M - 1 = 0.2$. One sees that the difference of the velocity distributions is appreciable even for such weak shock waves.

The author wishes to express her gratitude to Professor G. E. Uhlenbeck for his interest in this work and for his helpful suggestions and discussions.

REFERENCES

[1] C.S.Wang Chang, On the dispersion of sound in helium, Chap. II of this book.
[2] R.Becker, Z. Physik 8 (1922) 321-362.
[3] L.H.Thomas, J. Chem. Phys. 12 (1944) 449-453.
[4] C.S.Wang Chang and G. E. Uhlenbeck, On the transport phenomena in rarified gases, Chap. I of this book.

Chapter IV

ON THE PROPAGATION OF SOUND IN MONATOMIC GASES

C.S. WANG CHANG and G.E. UHLENBECK
University of Michigan (1952)

1. INTRODUCTION

Among the problems connected with the study of the transport phenomena in gases, the question of the sound propagation in monatomic gases is of special interest. The reason for this lies in the fact that by varying the ratio of the wavelength of sound, Λ, to the mean free path of the gas molecules, λ, one can study in the most direct way the transition from the so-called Clausius-gas to the Knudsen-gas regimes, which correspond to the limiting cases $\Lambda \gg \lambda$ and $\Lambda \ll \lambda$.

The classical theory of sound propagation is based on the Stokes-Navier equations, which lead to the dispersion law:

$$\omega^3 - \omega V_0^2 \sigma^2 - \frac{4i}{3}\frac{\mu}{\rho}\omega^2\sigma^2 - \frac{2i}{3}\frac{m\nu}{k\rho}\omega^2\sigma^2 + \frac{2i}{5}V_0^2\frac{m\nu}{k\rho}\sigma^4 - \frac{8}{9}\frac{m\mu\nu}{k\rho^2}\omega\sigma^4 = 0 , \qquad (1)$$

where ω is the frequency of sound, $\sigma = \sigma_1 - i\sigma_2$ is the complex wave number, V_0 is the velocity of sound at zero frequency, k is the Boltzmann constant, and m, ρ, μ, and ν are the molecular mass, the density, the viscosity coefficient, and the heat conduction of the gas respectively. One can solve eq. (1) exactly for the real and imaginary parts of σ, and thus obtain the "exact" expressions for the dispersion and absorption of sound. But for gases, it is in the spirit of the derivation of the Stokes-Navier equations from the Boltzmann equation to develop eq. (1) in powers of λ/Λ, and keep only terms up to the first order in this ratio. Doing this, one finds that there is no dispersion, the velocity of sound, V, is

$$V = V_0 ,$$

and the absorption coefficient σ_2 is given by:

$$\sigma_2 = \frac{\omega^2}{\rho V_0^3} \left(\frac{2}{3} \mu + \frac{2}{15} \frac{m\nu}{k} \right) .$$

For Maxwell molecules, i.e. molecules repelling with the Kr^{-5} force law, $\nu = \frac{15}{4} (k/m)\mu$, the expression reduces to

$$\sigma_2 = \frac{7}{6} \frac{\mu \omega^2}{\rho V_0^3} .$$

One can go to higher approximations by calculating from the Boltzmann equation higher-order corrections to the local Maxwell-Boltzmann distribution. From these one obtains the higher-order terms in the pressure tensor and heat conduction vector, leading to the corresponding higher-order hydrodynamic equations. From the second-order hydrodynamic equations (the Burnett equations) Primakoff [1] and later Tsien and Schamberg [2] have computed the dispersion and absorption of sound. Their results were in error on account of an error in the heat conduction vector given in Chapman's book [3]. This error was corrected by Chang and Uhlenbeck [4], who also carried the calculation one step further, i.e. computed the next term of the absorption coefficient. For Maxwell molecules the results are:

$$\sigma_1 = \frac{\omega}{V} = \frac{\omega}{V_0} \left\{ 1 - \frac{215}{72} \frac{\mu^2 \omega^2}{\rho^2 V_0^4} + \ldots \right\} , \tag{2a}$$

$$\sigma_2 = \frac{\mu \omega^2}{\rho V_0^3} \left\{ \frac{7}{6} - \frac{5155}{6.72} \frac{\mu^2 \omega^2}{\rho^2 V_0^4} + \ldots \right\} . \tag{2b}$$

Chang [5] has also reported results from the Burnett approximation for helium using for the force law a form derived from the second virial coefficient by de Boer and Michels. For room temperatures, the numerical coefficient 7/6 in eq. (2b) is replaced by 7.01/6, while the coefficient in eq. (2a) is replaced by 230.9/72, which indicates that the dependence of these coefficients on the molecular model is not very strong.

Recently measurements of the absorption and the dispersion of sound in helium have been made by Greenspan [6] using very low pressure and very high frequency. These experiments go beyond the range of validity of the Stokes-Navier and Burnett approximations, and probably they are beyond the range of convergence of any series expansion in λ/Λ. Thus, it seems desirable to make a further study of the theoretical aspect of the problem. To do this, it seems better to go back to the Boltzmann equation. In the Enskog-Chapman successive approximation method one does *not* distinguish between the *size* of the disturbance from the equilibrium state and the *scale* (compared to λ) of the disturbance. One concentrates on the derivation of the macroscopic equations of motion, which as a result are not linear. But for the sound propagation the nonlinear terms can always be neglected, since one can always assume that the intensity of the sound is sufficiently small. Thus the Enskog-Chapman method requires too much unnecessary work.

It is more practical to assume from the beginning that the size of the disturbance from the equilibrium state is small, and that the dependence on time and space coordinates is like $\exp i(\omega t - \sigma z)$ with *no* restrictions on the scale in time and space as compared to the time between collisions and the mean free path λ. It turns out that in this way it is easy to derive and extend results like eq. (2), and one may hope to obtain in this way the complete dispersion law for all values of λ/Λ. The same idea has occurred to Dr. Mott-Smith *, and his results are in substantial agreement with ours. In the following, we will give an account of our work. A joint publication with Dr. Mott-Smith is in preparation [7*].

2. THE BOLTZMANN EQUATION FOR A SMALL DISTURBANCE FROM EQUILIBRIUM

The Boltzmann equation for a monatomic gas when there is no outside force is

$$\frac{\mathrm{D}f}{\mathrm{D}t} \equiv \frac{\partial f}{\partial t} + \xi \frac{\partial f}{\partial x} + \eta \frac{\partial f}{\partial y} + \zeta \frac{\partial f}{\partial z}$$

$$= \iiint \mathrm{d}\xi_1 \, \mathrm{d}\eta_1 \, \mathrm{d}\zeta_1 \iint \mathrm{d}\epsilon \, \mathrm{d}\theta \, \sin\theta \, gI(g,\theta)(f' f_1' - f f_1) \,, \tag{3}$$

where $f(\xi, \eta, \zeta; x, y, z; t)$ is the velocity distribution function at time t and at the point (x, y, z), f_1 is the distribution function corresponding to the velocity components ξ_1, η_1, and ζ_1, and f' and f_1'

* We thank Dr. Mott-Smith for sending us his manuscript.

are the distribution functions corresponding to the velocities after collision, while $I(g,\theta)$ is the differential collision cross section corresponding to a turning of the relative velocity g over the angle θ into the solid angle $\sin\theta \, d\theta \, d\epsilon$. For the case of a small disturbance the "Ansatz" for f is:

$$f = f_0[1 + h(\xi, \eta, \zeta; x, y, z; t)] , \tag{4}$$

where f_0 is the *complete* equilibrium distribution function with no mass velocity * and $h \ll 1$, but no assumptions are made about the space and time variations of h. Substituting eq. (4) into (3) and introducing the dimensionless velocity

$$c = \sqrt{\frac{m}{2kT}} (\xi, \eta, \zeta)$$

we obtain, in the first approximation

$$\sqrt{\frac{m}{2kT}} \frac{\partial h}{\partial t} + c_x \frac{\partial h}{\partial x} + c_y \frac{\partial h}{\partial y} + c_z \frac{\partial h}{\partial z} = n J(h) , \tag{5}$$

where n is the number density and

$$J(h) \equiv \frac{1}{\pi^{\frac{3}{2}}} \iiint dc_1 e^{-c_1^2} \iint d\epsilon \, d\theta \, \sin\theta \cdot g I(g,\theta)(h' + h_1' - h - h_1) , \tag{6}$$

where g is also dimensionless. The operator J will be called the collision operator ‡. It has the dimension of an area and the order of magnitude of the collision cross section. Eq. (5) is a linear integral differential equation. For further discussion, it is convenient to consider first the eigenfunctions, ψ_i, of the collision operator J defined as

$$J(\psi_i) = \lambda_i \psi_i , \tag{7}$$

where λ_i is the eigenvalue. We will list some simple properties of the system of eigenfunctions and eigenvalues which are valid for any kind of interatomic force.

1) As a consequence of the conservation theorems of the number of particles, momentum, and energy during a collision, five of the eigenfunctions are known. They are, except for normalization factors:

$$\psi_{000} \propto 1 ,$$

* The density and the temperature are therefore supposed to be constant.
‡ It is to be noted that our definition of $J(h)$ differs from that of Chapman by a factor $-(m/2\pi kT)^{\frac{3}{2}} \sqrt{2kT/m} \, e^{-c^2}$.

$$\psi_{010} \propto c_z\,, \qquad \psi_{01-1} \propto c_x + ic_y\,, \qquad \psi_{011} \propto c_x - ic_y\,,$$

$$\psi_{100} \propto c^2 - \tfrac{3}{2}\,.$$

The corresponding eigenvalues are all zero. The meaning of the labels of the eigenfunctions will be explained below.

2) All the other eigenvalues are negative. This can be shown as follows: Multiplying both sides of eq. (7) by $\psi_i e^{-c^2}$ and integrating over dc,

$$\lambda_i = \frac{\iiint dc\, e^{-c^2}\psi_i J(\psi_i)}{\iiint dc\, e^{-c^2}(\psi_i)^2}\,.$$

The denominator is positive, while the numerator * can be transformed into

$$-\tfrac{1}{4}\int dc \int dc_1\, e^{-c^2-c_1^2} \int d\epsilon \int d\theta\, \sin\theta\, gI(g,\theta)$$
$$\times\, (\psi_i' + \psi_{1i}' - \psi_i - \psi_{1i})^2$$

and is therefore negative or zero. and zero only for the five values of ψ's mentioned above. Hence all the non-zero eigenvalues are negative.

3) Both Enskog [7] and Chapman have given the expression for J in the standard form, namely:

$$J(\psi) = -\sigma(c)\psi - \int dc_1\, e^{-c_1^2} K(c, c_1)\psi(c_1)\,, \tag{8}$$

where

$$\sigma(c) \equiv \frac{2}{\sqrt{\pi}}\int dc_1\, e^{-c_1^2}\int d\theta\, \sin\theta\, gI(g,\theta) \tag{9a}$$

and

$$K(c, c_1) = \frac{2}{\sqrt{\pi}}\int_0^\pi d\theta\, \sin\theta\, \{gI(g,\theta)$$
$$-\, g\csc^4(\tfrac{1}{2}\theta)\,[I(g\csc(\tfrac{1}{2}\theta),\theta) + I(g\csc(\tfrac{1}{2}\theta), \pi-\theta)] \tag{9b}$$
$$\times\, e^{-(c^2+c_1^2-2\,c\cdot c\ \ )\cot^2(\frac{1}{2}\theta)}\, J_0(-2icc_1 \sin\widehat{c\,c_1}\, \cot(\tfrac{1}{2}\theta))\}\,.$$

* It is essentially the negative of the bracket symbol $[\psi,\psi]$ used by Chapman.

One sees that $K(c, c_1)$ is symmetric in c and c_1 and is isotropic, which means that K depends only on the magnitudes of c and c_1 and on the angle, ϕ', between them.

4) From the isotropy of the kernel $K(c, c_1)$ it follows that the eigenfunctions ψ will have the form:

$$\psi_l(c) = f_l(c^2) Y_l(\phi, \chi) \,, \tag{10}$$

where ϕ and χ are the polar angles of c with respect to a set of fixed axes, and where the radial part $f_l(c^2)$ of the eigenfunction fulfills the equation:

$$-\sigma(c) f_l - \frac{4\pi}{2l+1} \int dc_1 c_1^2 \, e^{-c_1^2} f_l(c_1^2) \, G_l(c, c_1) = \lambda_l f_l \,. \tag{11}$$

$G_l(c, c_1)$ is defined by the development of $K(c, c_1, \cos\phi')$ in zonal harmonics:

$$K(c, c_1, \cos\phi') = \sum_{l'=0}^{\infty} G_{l'}(c, c_1) P_{l'}(\cos\phi') \,. \tag{12}$$

The proof is as follows: Introducing eq. (10) and eq. (12) into the integral in eq. (8), one gets:

$$\sum_{l'=0}^{\infty} \int d\boldsymbol{c}_1 \, e^{-c_1^2} G_{l'}(c, c_1) P_{l'}(\cos\phi') Y_l(\phi_1, \chi_1) f_l(c_1^2) \,,$$

where ϕ_1 and χ_1 are the polar angles of c_1 with respect to the same set of axes used for c. For the integration over c_1, we take the direction of c as the polar axis, so that the polar angles are ϕ' and χ', and $d\boldsymbol{c}_1 = c_1^2 dc_1 \sin\phi' \, d\phi' \, d\chi'$. The integration over χ' can then be performed, using

$$\int_0^{2\pi} d\chi' \, Y_l(\phi_1, \chi_1) = 2\pi P_l(\cos\phi') \, Y_l(\phi, \chi) \,,$$

which is a consequence of the addition theorem for spherical harmonics, and one obtains:

$$2\pi \sum_{l'=0}^{\infty} Y_l(\phi, \chi) \int_0^{\infty} \mathrm{d}c_1 c_1^2 \, \mathrm{e}^{-c_1^2} f_l(c_1^2) \, G_{l'}(c, c_1)$$

$$\times \int_0^{\pi} \mathrm{d}\phi' \sin\phi' \, P_{l'}(\cos\phi') \, P_l(\cos\phi')$$

$$= \frac{4\pi}{2l+1} Y_l(\phi, \chi) \int_0^{\infty} \mathrm{d}c_1 c_1^2 \, \mathrm{e}^{-c_1^2} f_l(c_1^2) G_l(c, c_1) \; .$$

Therefore if $f_l(c^2)$ fulfills eq. (11), $\psi_l(c)$ as given by eq. (10) will be an eigenfunction of $J(\psi)$.

5) The eigenvalues λ_l in eq. (11) will clearly depend on l, and on a "radial quantum number" r. We will write the eigenvalues as λ_{rl}; they are $(2l+1)$-fold degenerate, corresponding to the $2l+1$ different spherical harmonics $Y_l(\phi, \lambda)$ of order l. The radial eigenfunctions $f_{rl}(c^2)$ form an orthogonal set of functions with a weight factor $c^2 \, \mathrm{e}^{-c^2}$. The complete eigenfunctions will be written as

$$\psi_{rlm}(c) = N_{rlm} \, Y_l^m(\phi, \chi) f_{rl}(c^2) \; ,$$

where N_{rlm} are appropriate normalization constants, so that:

$$\int \mathrm{d}c \, \mathrm{e}^{-c^2} \, \psi_{rlm} \, \psi_{r'l'm'} = \delta_{rr'} \, \delta_{ll'} \, \delta_{mm'} \; .$$

The simplest problem to consider is the case where the disturbance h depends only on t, but not on x, y, and z. Then, eq. (5) becomes

$$\frac{\partial h}{\partial t} = n \sqrt{\frac{2kT}{m}} \, J(h) \; . \tag{13}$$

Developing h in terms of the set of normalized eigenfunctions of J,

$$h = \sum_0^{\infty} \alpha_{r'l'm'} \, \psi_{r'l'm'}(c) \; .$$

Sunstituting in eq. (13), multiplying by $\psi_{rlm} \, \mathrm{e}^{-c^2}$, and integrating over $\mathrm{d}c$, one obtains the set of simple differential equations:

$$\frac{\mathrm{d}\alpha_{rlm}}{\mathrm{d}t} = n \sqrt{\frac{2kT}{m}} \, \lambda_{rl} \, \alpha_{rlm} \tag{14}$$

for all values of n, l, and m. The solutions are

$$\alpha_{rlm}(t) = \alpha_{rlm}(0)\, e^{n \sqrt{\frac{2kT}{m}} \lambda_{rl} t} .$$

For all non-zero eigenvalues, the coefficients α_{rlm} approach zero exponentially. For the five zero eigenvalues, the α's remain constant. They are determined by the initial values of the density, mass velocity and temperature. The solution is clearly a special case of the general result of Boltzmann (proved by means of the H-theorem) that the equilibrium distribution, as described by a Maxwell distribution with a mass velocity, is reached monotonically in time.

For the problem of sound propagation we will assume h to be of the form:

$$h(\xi, \eta, \zeta)\, e^{i(\omega t - \sigma z)}$$

with ω real and $\sigma = \sigma_1 - i\sigma_2$. Eq. (5) now becomes:

$$i\left(\sqrt{\frac{m}{2kT}}\,\omega - \sigma c_z\right) h = nJ(h) . \tag{15}$$

Developing h as before, one obtains an infinite set of homogeneous linear equations with coefficients α_{rlm}. In order that they have a solution, the infinite determinant must be zero. This gives a relation between ω and σ, which is an expression of the exact dispersion law for the gas.

For the further development we will limit ourselves to the case of Maxwell molecules. For this case we have found an explicit expression for the "radial" eigenfunctions f_{rl}, which leads to a straightforward evaluation of the coefficients in the series expansions in eq. (2).

3. PROPAGATION OF SOUND IN A GAS OF MAXWELL MOLECULES [8*]

For Maxwell molecules the collision probability per second is independent of the relative velocity g. It is convenient to write, if g is dimensionless,

$$gI(g, \theta) = \sqrt{\frac{\kappa}{kT}}\, F(\theta) ,$$

or $gI(g,\theta) = \sqrt{2\kappa/m}\, F(\theta)$ if g is not dimensionless, and redefine the collision operator as:

$$J(h) = \frac{1}{\pi^{\frac{3}{2}}} \int \mathrm{d}c_1 \, e^{-c_1^2} \int \mathrm{d}\epsilon \int \mathrm{d}\theta \, \sin\theta \, F(\theta) \, (h' + h_1' - h - h_1) \, , \tag{17}$$

where κ is the force constant and $F(\theta)$ is dimensionless and a function of θ only, so that the collision operator J is also dimensionless and the eigenvalues of J will be pure numbers. In appendix B, the definition and some of the properties of $F(\theta)$ will be given.

Adopting these changes, eq. (15) becomes

$$i\left(\omega - \sqrt{\frac{2kT}{m}}\, \sigma c_z\right) h = n \sqrt{\frac{2\kappa}{m}}\, J(h) \, , \tag{18}$$

where

$$J(h) = - A_0 h - \int \mathrm{d}c_1 \, e^{-c_1^2} h(c_1) K(c, c_1, \cos\phi') \tag{19}$$

with

$$A_0 \equiv 2\pi \int_0^{\pi} \mathrm{d}\theta \, \sin\theta \, F(\theta)$$

and

$$K(c, c_1, \cos\phi') = \frac{1}{\pi^{\frac{3}{2}}} \Bigg\{ A_0 - 4\pi \int_0^{\pi} \mathrm{d}\theta \, \cos(\tfrac{1}{2}\theta) \csc^2(\tfrac{1}{2}\theta) \left[F(\theta) + F(\pi - \theta)\right]$$

$$\times \, e^{-(c^2 + c_1^2 - 2 c \cdot c \;\;) \cot^2(\frac{1}{2}\theta)} J_0(-2icc_1 \sin\phi' \cot(\tfrac{1}{2}\theta)) \Bigg\} \, . \tag{20}$$

Since for small θ, $F(\theta)$ becomes infinite as $\theta^{-\frac{5}{2}}$ (see appendix B), both terms in $J(h)$ are really infinite. However, the combination of the two terms is finite.

To find the radial eigenfunctions one first has to determine $G_l(c, c_1)$. From eq. (12), we have:

$$G_l = \frac{2l+1}{2} \int_0^{\pi} \mathrm{d}\phi' \sin\phi' \, P_l(\cos\phi') K(c, c_1, \cos\phi') \, .$$

With the help of the integral formula [8]

$$\int_0^{\pi} \mathrm{d}\theta \, \sin^{\nu+\frac{1}{2}}\theta \; e^{iz\cos\theta\cos\psi} J_{\nu-\frac{1}{2}}(z \sin\theta \sin\psi) C_\gamma^\nu(\cos\theta)$$

$$= \sqrt{\frac{2\pi}{z}} \, i^\gamma \sin^{\nu-\frac{1}{2}}\psi \, C_\gamma^\nu(\cos\psi) \, J_{\nu+\gamma}(z) \, .$$

the ϕ' integration can be performed, leading to:

$$G_l(c, c_1) = \frac{1}{\pi^{\frac{3}{2}}} A_0 \delta_{l0} - \frac{2(2l+1)}{\sqrt{cc_1}} (i)^{l+\frac{1}{2}} \int_0^\pi d\theta \sqrt{\cos(\frac{1}{2}\theta)}\, P_l(\cos(\frac{1}{2}\theta))$$

$$\times [F(\theta) + F(\pi-\theta)]\, e^{-(c^2+c_1^2)\cot^2(\frac{1}{2}\theta)} J_{l+\frac{1}{2}}(-2icc_1 \cot(\frac{1}{2}\theta) \csc(\frac{1}{2}\theta)) \quad (21)$$

and the "radial" integral equation:

$$-A_0 f_l(c^2) - \frac{4}{\sqrt{\pi}} A_0 \delta_{l0} \int_0^\infty dc_1 c_1^2 e^{-c_1^2} f_l(c_1^2)$$

$$+ \frac{8\pi}{\sqrt{cc_1}} (i)^{l+\frac{1}{2}} \int_0^\infty dc_1 c_1^2 e^{-c^2} f_l(c_1^2) \int_0^\pi d\theta \sqrt{\cos(\frac{1}{2}\theta)} \csc(\frac{1}{2}\theta)$$

$$\times P_l(\cos(\frac{1}{2}\theta)) [F(\theta) + F(\pi-\theta)]\, e^{-(c^2+c_1^2)\cot^2(\frac{1}{2}\theta)}$$

$$\times J_{l+\frac{1}{2}}(-2icc_1 \cot(\frac{1}{2}\theta) \csc(\frac{1}{2}\theta)) = \lambda_l f_l(c^2) \,. \quad (22)$$

We assert that the eigenfunctions are:

$$f_{rl}(c^2) = c^l S_{l+\frac{1}{2}}^{(r)}(c^2) \equiv c^l \sum_p (-)^p c^{2p} \frac{(l+\frac{1}{2}+r)!}{p!(r-p)!(l+\frac{1}{2}+p)!} \,, \quad (23)$$

where $S_{l+\frac{1}{2}}^{(r)}(c^2)$ are the Sonine polynomials of degree r and order $l + \frac{1}{2}$, which are defined by the last part of eq. (23). For some of the properties of these polynomials, see appendix B. To verify the above statement and to find the eigenvalues we substitute eq. (23) into eq. (22). The second term of eq. (22) is zero unless l is zero, and because of the orthogonality of the Sonine polynomials r must also be zero. Since f_{00} is 1, the second term can be written as $-A_0\delta_{r0}\delta_{l0}f_{rl}$. Hence, interchanging the order of integration in the last term, one finds:

$$-A_0 f_{rl}(c^2) - A_0 \delta_{r0} \delta_{l0} f_{rl}(c^2)$$

$$+ 8\pi(i)^{l+\frac{1}{2}} \frac{1}{\sqrt{c}} \sum_p (-)^p \int_0^\pi d\theta \sqrt{\cos(\frac{1}{2}\theta)} \csc(\frac{1}{2}\theta)\, P_l(\cos(\frac{1}{2}\theta))\, e^{-c^2\cot^2(\frac{1}{2}\theta)}$$

$$\times [F(\theta) + F(\pi-\theta)] \frac{(l+\frac{1}{2}+r)!}{p!(r-p)!(l+\frac{1}{2}+p)!}$$

$$\times \int_0^\infty dc_1 c_1^{l+\frac{3}{2}+2p} e^{-c_1^2\csc^2(\frac{1}{2}\theta)} J_{l+\frac{1}{2}}(-2icc_1 \cot(\frac{1}{2}\theta) \csc(\frac{1}{2}\theta)).$$

The c_1 integral is evaluated by using the formula [9]:

$$\int_0^\infty \mathrm{d}t\, t^{\mu-1}\, \mathrm{e}^{-p^2t^2}\, J_\nu(at) = \frac{\Gamma(\frac{1}{2}(\mu+\nu))(a/2p)^\nu}{2p^\mu\,\Gamma(\nu+1)}\, \mathrm{e}^{-a^2/4p^2}\; {}_1F_1\left(\frac{\nu-\mu}{2}+1,\, \nu+1,\, \frac{a^2}{4p^2}\right),$$

and one obtains:

$$-A_0 f_{rl}(c^2)(1+\delta_{r0}\delta_{l0}) + \frac{2\pi}{(l+\frac{1}{2})!}\, c^l \int_0^\pi \mathrm{d}\theta\, \sin\theta\, \cos^l(\tfrac{1}{2}\theta)$$

$$\times P_l(\cos(\tfrac{1}{2}\theta))\,[F(\theta)+F(\pi-\theta)] \sum_{p=0} (-)^p \frac{(l+\frac{1}{2}+r)!}{p!(r-p)!}$$

$$\times \sin^{2p}(\tfrac{1}{2}\theta)\; {}_1F_1(-p,\, l+\tfrac{3}{2},\, -c^2\cot^2(\tfrac{1}{2}\theta))$$

$$= -A_0 f_{rl}(c^2)(1+\delta_{r0}\delta_{l0}) + 2\pi c^l \int_0^\pi \mathrm{d}\theta\, \sin\theta\, \cos^l(\tfrac{1}{2}\theta))$$

$$\times P_l(\cos(\tfrac{1}{2}\theta))[F(\theta)+F(\pi-\theta)] \sum_{p=0} (-)^p \frac{(l+\frac{1}{2}+r)!}{p!(r-p)!} \sin^{2p}(\tfrac{1}{2}\theta))$$

$$\times \sum_{j=0} \frac{p!}{j!(p-j)!(l+\frac{1}{2}+j)!}\, c^{2j} \cos^{2j}(\tfrac{1}{2}\theta)\, \sin^{-2j}(\tfrac{1}{2}\theta)\ .$$

Interchanging the order of summations, the sum over p is seen to be just $(-)^j \cos^{2(r-j)}(\frac{1}{2}\theta)$; the sum over j is then:

$$\sum_{j=0}^{r} (-)^j c^{2j} \frac{(l+\frac{1}{2}+r)!}{j!(r-j)!(l+\frac{1}{2}+j)!} = S^{(r)}_{l+\frac{1}{2}}(c^2)\ .$$

We thus arrive at the desired result; the eigenvalues are given by:

$$\lambda_{rl} = 2\pi \int_0^\pi \mathrm{d}\theta\, \sin\theta\, F(\theta) \qquad (24)$$

$$\times [\cos^{2r+l}(\tfrac{1}{2}\theta)\, P_l(\cos(\tfrac{1}{2}\theta)) + \sin^{2r+l}(\tfrac{1}{2}\theta)\, P_l(\sin(\tfrac{1}{2}\theta)) - (1+\delta_{r0}\delta_{l0})]\ .$$

All eigenvalues are negative except λ_{00}, λ_{10}, and λ_{01}, which are zero as in the general case. For the same value of l the eigenvalues decrease with increasing values of r. It can be proved that these is *no* lower bound (see appendix C). A few of the eigenvalues are listed in table 1. Here $A_{2k} = 2\pi \int_0^\pi \mathrm{d}\theta\, \sin^{2k+1}\theta\; F(\theta)$.

Table 1
A partial list of the eigenvalues of the collision operator for the Maxwell molecules

l \ r	0	1	2
0	0	0	$-\frac{1}{2}A_2$
1	0	$-\frac{1}{2}A_2$	$-\frac{3}{4}A_2$
2	$-\frac{3}{4}A_2$	$-\frac{7}{8}A_2$	$-\frac{9}{8}A_2+\frac{3}{16}A_4$
3	$-\frac{9}{8}A_2$	$-\frac{11}{8}A_2+\frac{5}{16}A_4$	$-\frac{13}{8}A_2+\frac{19}{32}A_4$
4	$-\frac{7}{4}A_2+\frac{35}{64}A_4$	$-\frac{16}{8}A_2+\frac{115}{128}A_4$	$-\frac{18}{8}A_2+\frac{171}{128}A_4-\frac{35}{256}A_6$

For the problem of the propagation of sound we can take for the complete set of eigenfunctions:

$$\psi_{rl} = \sqrt{\frac{r!(l+\frac{1}{2})}{\pi(l+\frac{1}{2}+r)!}}\, c^l P_l(\cos\phi)\, S^{(r)}_{l+\frac{1}{2}}(c^2) .$$

The ψ_{rl} are normalized to unity with the weight factor e^{-c^2}. They are polynomials in c_x, c_y, and c_z of degree $2r+l$. And the first eight are listed in table 2. Developing the disturbance h in the ψ_{rl},

Table 2
First eight of the eigenfunctions of the collision operator for the Maxwell molecules

ψ_{00}	$\pi^{-3/4}$
ψ_{01}	$\sqrt{2}\,\pi^{-3/4}c_z$
ψ_{10}	$\sqrt{\frac{2}{3}}\,\pi^{-3/4}(\frac{3}{2}-c^2)$
ψ_{02}	$\frac{1}{\sqrt{3}}\,\pi^{-3/4}(3c_z^2-c^2)$
ψ_{11}	$\frac{2}{\sqrt{5}}\,\pi^{-3/4}c_z(\frac{5}{2}-c^2)$
ψ_{03}	$\sqrt{\frac{2}{15}}\,\pi^{-3/4}(5c_z^3-3c^2c_z)$
ψ_{12}	$\sqrt{\frac{2}{21}}\,\pi^{-3/4}(3c_z^2-c^2)(\frac{7}{2}-c^2)$
ψ_{20}	$\sqrt{\frac{2}{15}}\,\pi^{-3/4}(\frac{15}{4}-5c^2+c^4)$

one gets for the coefficients α_{rl} the set of equations:

$$i\omega\alpha_{rl} - i\sqrt{\frac{2kT}{m}}\,\sigma \sum_{r'l'} M_{rl,r'l'}\alpha_{r'l'} = n\sqrt{\frac{2\kappa}{m}}\,\alpha_{rl}\lambda_{rl}\,, \qquad (25)$$

where

$$M_{rl,r'l'} = \int \mathrm{d}c\, \mathrm{e}^{-c^2} c_z \psi_{rl}\psi_{r'l'}$$

$M_{rl,r'l'}$ is symmetrical with respect to the pair of indices rl and $r'l'$. Using the recurrence relation

$$\cos\theta\, P_l(\cos\theta) = \frac{1}{2l+1}\left[(l+1)P_{l+1} + lP_{l-1}\right]$$

and the integral properties of the Sonine polynomial given in appendix C, one deduces easily the selection rules for the matrix elements $M_{rl,r'l'}$, namely

$$M_{rl,r'l'} = 0$$

unless $l' = l \pm 1$ and $2r' + l' = 2r + l \pm 1$. The complete expression for M is:

$$M_{rl,r'l'} = (l+1)\sqrt{\frac{(r+l+\frac{3}{2})}{(2l+1)(2l+3)}}\,\delta_{rr'}\delta_{l+1,l'}$$

$$- (l+1)\sqrt{\frac{r}{(2l+1)(2l+3)}}\,\delta_{r-1,r'}\delta_{l+1,l'} + l\sqrt{\frac{r+l+\frac{1}{2}}{(2l-1)(2l+1)}}\,\delta_{rr'}\delta_{l-1,l'}$$

$$- l\sqrt{\frac{r+1}{(2l-1)(2l+1)}}\,\delta_{r+1,r'}\delta_{l-1,l'}\,.$$

The infinite set of linear equations (25) will have a solution if their determinant is zero. This will give a relation between ω and σ which is the general dispersion law of the gas.

We have not succeeded in developing a general discussion of the dispersion law, and we have therefore gone back to a successive-approximation method analogous to the Enskog-Chapman development. To do this, we have to choose a special ordering of the linear equations and consequently of the infinite determinant. The most natural ordering is according to the degree $2r + l$ of the eigenfunctions ψ_{rl} and for each group according to increasing values of l. The first eleven rows and columns of the determinant ordered in this way are shown in the page following. We have further introduced the dimensionless quantities

$$\omega_0 = \frac{\omega}{n\sqrt{\dfrac{2\kappa}{m}}}, \qquad \sigma_0 = \sqrt{\frac{kT}{\kappa}}\,\frac{\sigma}{n},$$

so that

$$\frac{\sigma_0}{\omega_0} = \sqrt{\frac{2kT}{m}}\,\frac{\sigma}{\omega}.$$

The heavy lines indicate the successive stages of approximation, the choice of which will be explained in the next section.

2r+l	l	r	0	1	2		3		4			5	
		l	0	1	0	2	1	3	0	2	4	1	3
		r	0	0	1	0	1	0	2	0	0	2	1
0	0	0	ω_0	$-\frac{1}{\sqrt{2}}\sigma_0$	0	0	0	0	0	0	0	0	0
1	1	0	$-\frac{1}{\sqrt{2}}\sigma_0$	ω_0	$\frac{1}{\sqrt{3}}\sigma_0$	$-\sqrt{\frac{2}{3}}\sigma_0$	0	0	0	0	0	0	0
2	0	1	0	$\frac{1}{\sqrt{3}}\sigma_0$	ω_0	0	$-\sqrt{\frac{5}{6}}\sigma_0$	0	0	0	0	0	0
	2	0	0	$-\sqrt{\frac{2}{3}}\sigma_0$	0	$\omega_0+i\lambda_{02}$	$\frac{2}{\sqrt{15}}\sigma_0$	$-\frac{3}{\sqrt{10}}\sigma_0$	0	0	0	0	0
3	1	1	0	0	$-\sqrt{\frac{5}{6}}\sigma_0$	$\frac{2}{\sqrt{15}}\sigma_0$	$\omega_0+i\lambda_{11}$	0	$\sqrt{\frac{2}{3}}\sigma_0$	$-\sqrt{\frac{14}{15}}\sigma_0$	0	0	0
	3	0	0	0	0	$-\frac{3}{\sqrt{10}}\sigma_0$	0	$\omega_0+i\lambda_{03}$	0	$\frac{3}{\sqrt{35}}\sigma_0$	$-\sqrt{\frac{8}{7}}\sigma_0$	0	0
4	0	2	0	0	0	0	$\sqrt{\frac{2}{3}}\sigma_0$	0	$\omega_0+i\lambda_{20}$	0	0	$-\sqrt{\frac{7}{6}}\sigma_0$	0
	2	1	0	0	0	0	$-\sqrt{\frac{14}{15}}\sigma_0$	$\frac{3}{\sqrt{35}}\sigma_0$	0	$\omega_0+i\lambda_{12}$	0	$\sqrt{\frac{8}{15}}\sigma_0$	$-\frac{9}{\sqrt{70}}\sigma_0$
	4	0	0	0	0	0	0	$-\sqrt{\frac{8}{7}}\sigma_0$	0	0	$\omega_0+i\lambda_{04}$	0	$\frac{4}{\sqrt{63}}\sigma_0$
5	1	2	0	0	0	0	0	0	$-\sqrt{\frac{7}{6}}\sigma_0$	$\sqrt{\frac{8}{15}}\sigma_0$	0	$\omega_0+i\lambda_{21}$	0
	3	1	0	0	0	0	0	0	0	$-\frac{9}{\sqrt{70}}\sigma_0$	$\frac{4}{\sqrt{63}}\sigma_0$	0	$\omega_0+i\lambda_{13}$

4. SUCCESSIVE APPROXIMATIONS OF THE DISPERSION LAW FOR A MAXWELL GAS

As in the Enskog development, we will try to construct a successive-approximation method which will give the dispersion law as a series in λ/Λ, the ratio of the mean free path to the wavelength of the sound. It is to be noted that in the determinant the λ_{rl} are of the order unity, while ω_0 and σ_0 are small, of the order λ/Λ.

4.1. *Zeroth approximation ("ideal fluid")*

Clearly, taking only the first three rows and columns of the determinant, one uses only the zero eigenvalues of the collision operation $J(h)$. These correspond to the basic conservation laws of number, momentum, and energy, and one must expect therefore that this approximation corresponds to the ideal fluid approximation in the usual theory. One gets in fact,

$$\Delta_3 = \begin{vmatrix} \omega_0 & -\frac{\sigma_0}{\sqrt{2}} & 0 \\ -\frac{\sigma_0}{\sqrt{2}} & \omega_0 & \frac{\sigma_0}{\sqrt{3}} \\ 0 & \frac{\sigma_0}{\sqrt{3}} & \omega_0 \end{vmatrix} = \omega_0(\omega_0^2 - \tfrac{5}{6}\sigma_0^2) = 0$$

of which the roots are

$$\omega_0 = 0 ,$$

and

$$\omega_0^2 = \tfrac{5}{6}\sigma_0^2 \qquad \text{or} \qquad \omega = V_0\sigma ,$$

with

$$V_0 = \sqrt{\frac{5kT}{3m}} ,$$

which is the well-known result. One can also determine the zeroth approximation of the first three expansion coefficients α_{rl}. One of them is arbitrary, which corresponds clearly to the arbitrariness of the amplitude of the sound. One finds for the root $\omega_0 = \sqrt{\frac{5}{6}}\,\sigma_0$:

$$h = \alpha_{00}\left\{1 + \sqrt{\frac{5m}{3kT}}\,\xi - \frac{2}{3}\left(\frac{3}{2} - \frac{m\xi^2}{2kT}\right) \dots \quad . \right\} \tag{26}$$

By computing the number density, the mass velocity, and the temperature, it can easily be verified that eq. (26) is of the form to be expected.

4.2. *First approximation* ("*Stokes-Navier*")

For this approximation we add two rows and two columns to the determinant. A comparison with the Enskog development shows that this corresponds to the Stokes-Navier approximation. Calling the determinant Δ_5 and developing, one obtains

$$\Delta_5 = (\omega_0 + i\lambda_{02})(\omega_0 + i\lambda_{11})\Delta_3 - \tfrac{2}{3}\,\omega_0^2\sigma_0^2(\omega_0 + i\lambda_{11})$$

$$- \tfrac{5}{6}\,\sigma_0^2(\omega_0^2 - \tfrac{1}{2}\sigma_0^2)(\omega_0 + i\lambda_{02}) - \tfrac{4}{15}\omega_0^3\sigma_0^2 + \tfrac{1}{3}\,\omega_0\sigma_0^4\,, \tag{27}$$

where $\lambda_{02} = -\frac{3}{4}A_2$ and $\lambda_{11} = -\frac{1}{2}A_2$ as given in table 1. We are mainly interested in the development around the zeroth-order root $\sigma_0^2 = \frac{6}{5}\omega_0^2$, so we put

$$\sigma_0^2 = \frac{6\omega_0^2}{5} + \frac{b_1}{A_2} + \frac{b_2}{A_2^2} + \ldots$$

as a solution of $\Delta_5 = 0$ and determine b_1 and b_2 by equating coefficients of equal powers of A_2. This yields

$$b_1 = -\frac{56\,i\omega_0^3}{25}\,, \qquad b_2 = -\frac{64 \times 11\,\omega_0^4}{125}\,,$$

so that

$$\sigma_0^2 = \frac{6\omega_0^2}{5}\left(1 - \frac{28}{15}\,\frac{i\omega_0}{A_2} - \frac{2^5 \times 11}{3 \times 5^2}\,\frac{\omega_0^2}{A_2^2} + \ldots\right). \tag{28}$$

Changing to ordinary units and making use of the connection between A_2 and the viscosity coefficient, μ,

$$\frac{1}{n\sqrt{\frac{2\kappa}{m}}\,A_2} = \frac{5\mu}{4\rho V_0^2}\,,$$

it follows that

$$\sigma^2 = \frac{\omega^2}{V_0^2} - \frac{7i\mu\omega^3}{3\rho V_0^4} - \frac{22}{3}\,\frac{\mu^2\omega^4}{\rho^2 V_0^6} + \ldots\,. \tag{29}$$

Eq. (29) gives the following expressions for the velocity and coefficient of absorption of sound:

$$\frac{\omega}{V} = \frac{\omega}{V_0}\left\{1 - \frac{215}{72}\,\frac{\mu^2\omega^2}{\rho^2 V_0^4}\right\}, \tag{29a}$$

$$\sigma_2 = \frac{7\mu\omega^2}{6\rho V_0^3}\,. \tag{29b}$$

The first five expansion coefficients can again be determined in terms of α_{00} which remains arbitrary. For the root given by eq. (29) and up to the second power of μ, one gets:

$$\alpha_{01} = \sqrt{\frac{5}{3}}\,\alpha_{00}\left\{1 + \frac{7}{6}\,\frac{i\mu\omega}{\rho V_0^2} + \frac{13}{8}\left(\frac{\mu\omega}{\rho V_0^2}\right)^2 + \ldots\right\},$$

$$\alpha_{10} = -\sqrt{\frac{2}{3}}\,\alpha_{00}\left\{1 + \frac{5}{2}\,\frac{i\mu\omega}{\rho V_0^2} + \frac{5}{2}\left(\frac{\mu\omega}{\rho V_0^2}\right)^2 + \ldots\right\},$$

$$\alpha_{02} = \frac{10\sqrt{3}}{9}\,\alpha_{00}\,\frac{i\mu\omega}{\rho V_0^2}\left\{1 - \frac{2}{3}\,\frac{i\mu\omega}{\rho V_0^2} + \ldots\right\},$$

$$\alpha_{11} = -\frac{5\sqrt{6}}{6}\,\alpha_{00}\,\frac{i\mu\omega}{\rho V_0^2}\left\{1 - \frac{1}{6}\,\frac{i\mu\omega}{\rho V_0^2} + \ldots\right\}. \tag{30}$$

It is to be noted that the dispersion law, eq. (20) or eqs. (29a) and (29b), agrees with the result obtained from the Stokes-Navier equations (sect. 1) only up to the *first* order in μ. If the Stokes-Navier equations had been considered as exact and if they had been developed in powers of μ (specializing for the Maxwell gas for which $\nu = (\frac{15}{4})(k/m)\mu$, the velocity would have been obtained as

$$\frac{\omega}{V} = \frac{\omega}{V_0}\left\{1 - \frac{141}{72}\left(\frac{\mu\omega}{\rho V_0^2}\right)^2 + \ldots\right\}.$$

Our first approximation is therefore not quite identical with the Stokes-Navier approximation in the Enskog development. That the coefficient 215/72 in eq. (29a) is the right one *, follows from the second approximation, discussed below.

* It follows also from the Enskog development but only from the next order or the Burnett approximation, as noted in ref. [4].

4.3. *Second approximation* ("*Burnett*")

Three rows and columns are added which is again suggested by the Enskog development, giving:

$$\Delta_8 = \left\{(\omega_0+i\lambda_{03})(\omega_0+i\lambda_{20})(\omega_0+i\lambda_{12}) - \frac{9\sigma_0^2}{35}(\omega_0+i\lambda_{20})\right\}\Delta_5$$

$$-\sigma_0^2\{\tfrac{14}{15}(\omega_0+i\lambda_{02})(\omega_0+i\lambda_{03})(\omega_0+i\lambda_{20}) + \tfrac{2}{3}(\omega_0+i\lambda_{02})(\omega_0+i\lambda_{03})(\omega_0+i\lambda_{1}$$

$$+\tfrac{9}{10}(\omega_0+i\lambda_{11})(\omega_0+i\lambda_{20})(\omega_0+i\lambda_{12})\}\Delta_3$$

$$+\sigma_0^4\{\tfrac{28}{45}\omega_0^2(\omega_0+i\lambda_{20})(\omega_0+i\lambda_{03}) + \tfrac{4}{9}\omega_0^2(\omega_0+i\lambda_{03})(\omega_0+i\lambda_{12})$$

$$+\tfrac{3}{4}(\omega_0^2 - \tfrac{1}{2}\sigma_0^2)(\omega_0+i\lambda_{20})(\omega_0+i\lambda_{12})\}$$

$$+\sigma_0^4\{\tfrac{6}{35}(\omega_0+i\lambda_{02}) + \tfrac{9}{25}(\omega_0+i\lambda_{20}) + \tfrac{3}{5}(\omega_0+i\lambda_{12})\}\Delta_3$$

$$-\tfrac{2}{5}\omega_0\sigma_0^6(\omega_0+i\lambda_{20}) - \tfrac{4}{35}\omega_0^2\sigma_0^6 = 0\ .$$

Using the values of λ_{rl} given in table 1, and making the same kind of expansion for σ_0^2, but now up to A_2^{-4}, one gets:

$$\sigma_0^2 = \frac{6\omega_0^2}{5}\left\{1 - \frac{28}{15}\frac{i\omega_0}{A_2} - \frac{2^5\times 11}{3\times 5^2}\frac{\omega_0^2}{A_2^2} + \frac{2^5\times 37}{3\times 5^2}\frac{i\omega_0^3}{A_2^3} + \frac{2^7\times 56761}{3^3\times 5^4\times 7}\frac{\omega_0^4}{A_2^4}\cdots\right\}\ . \tag{31}$$

The expressions for the velocity and the absorption coefficient, which follow from this equation, are

$$\frac{\omega}{V} = \frac{\omega}{V_0}\left\{1 - \frac{215}{72}\frac{\mu^2\omega^2}{\rho^2 V_0^4} + \frac{4115101}{2^7\times 3^4\times 7}\frac{\mu^4\omega^4}{\rho^4 V_0^8}\cdots\right\}$$

$$= \frac{\omega}{V_0}\left\{1 - 2.99\frac{\mu^2\omega^2}{\rho^2 V_0^4} + 56.70\frac{\mu^4\omega^4}{\rho^4 V_0^8}\cdots\right\} \tag{31a}$$

and

$$\sigma_2 = \frac{\mu\omega^3}{\rho V_0^3}\left\{\frac{7}{6} - \frac{5155}{6 \times 72}\frac{\mu^2\omega^2}{\rho^2 V_0^4}\cdots\right\}$$

$$= \frac{\mu\omega^2}{\rho V_0^3}\left\{1.17 - 11.93\frac{\mu^2\omega^2}{\rho^2 V_0^4}\cdots\right\} . \qquad (31b)$$

The corresponding expansion coefficients α_{rl} are given as follows:

$$\alpha_{01} = \sqrt{\frac{5}{3}}\,\alpha_{00}\left\{1 + \frac{7}{6}\frac{i\mu\omega}{\rho V_0^2} + \frac{13}{8}\left(\frac{\mu\omega}{\rho V_0^2}\right)^2 - \frac{2831}{2^4 \times 3^3}\,i\left(\frac{\mu\omega}{\rho V_0^2}\right)^3 - \frac{5^2 \times 87907}{2^7 \times 3^4 \times 7}\left(\frac{\mu\omega}{\rho V_0^2}\right)^4\cdots\right\} ,$$

$$\alpha_{10} = -\sqrt{\frac{2}{3}}\,\alpha_{00}\left\{1 + \frac{5}{2}\frac{i\mu\omega}{\rho V_0^2} + \frac{5}{2}\left(\frac{\mu\omega}{\rho V_0^2}\right)^2 - \frac{5 \times 13}{4}\,i\left(\frac{\mu\omega}{\rho V_0^2}\right)^3 - \frac{5 \times 9 \times 59}{2^2 \times 7}\left(\frac{\mu\omega}{\rho V_0^2}\right)^4\cdots\right\} ,$$

$$\alpha_{02} = \frac{10\sqrt{3}}{9}\frac{i\mu\omega}{\rho V_0^2}\,\alpha_{00}\left\{1 - \frac{2}{3}\frac{i\mu\omega}{\rho V_0^2} - \frac{19}{9}\left(\frac{\mu\omega}{\rho V_0^2}\right)^2 + \frac{1333}{2 \times 3^3 \times 7}\,i\left(\frac{\mu\omega}{\rho V_0^2}\right)^3\cdots\right\} ,$$

$$\alpha_{11} = -\frac{5\sqrt{6}}{6}\frac{i\mu\omega}{\rho V_0^2}\,\alpha_{00}\left\{1 - \frac{1}{6}\frac{i\mu\omega}{\rho V_0^2} - \frac{89}{24}\left(\frac{\mu\omega}{\rho V_0^2}\right)^2 + \frac{67033}{2^4 \times 3^3 \times 7}\,i\left(\frac{\mu\omega}{\rho V_0^2}\right)^3\cdots\right\} ,$$

$$\alpha_{03} = -\frac{20}{9}\left(\frac{\mu\omega}{\rho V_0^2}\right)^2 \alpha_{00}\left\{1 - \frac{263}{2\times 3^2 \times 7}\, i\, \frac{\mu\omega}{\rho V_0^2}\cdots\right\},$$

$$\alpha_{20} = -\frac{25}{\sqrt{30}}\left(\frac{\mu\omega}{\rho V_0^2}\right)^2 \alpha_{00}\left\{1 - \frac{7}{2}\frac{i\mu\omega}{\rho V_0^2}\cdots\right\},$$

$$\alpha_{12} = \frac{10}{7}\sqrt{\frac{14}{3}}\left(\frac{\mu\omega}{\rho V_0^2}\right)^2 \alpha_{00}\left\{1 - \frac{13}{7}\frac{i\mu\omega}{\rho V_0^2}\cdots\right\}.$$

One can go on in this way. In the third approximation eleven rows and columns must be taken, which permits development up to terms of the order μ^6. The calculation becomes lengthy, and since the coefficients in the series increase rather rapidly so that the range of applicability of such a development seems rather limited, we have not gone any further. Several points should be noted in connection with the development outlined:

1) When the calculation is performed as outlined above, at no stage of the approximation are the results of the earlier approximations changed; each new stage simple adds two higher-order terms. This is seen by comparing eqs. (28) and (31), and it can be proved in general *. It is due to the fact that in the nth approximation the added rows and columns have no non-vanishing elements extending farther than the blocks belonging to the $(n-1)$th approximation. This in turn is due to the selection rule

$$2r' + l' = 2r + l \pm 1$$

in the matrix element $M_{rl,r'l'}$. This is also the reason why one gets at the nth stage of approximation a result for the dispersion law which is n steps better than the result obtained from the corresponding stage of approximation in the Enskog development, as we noted previously, in the Stokes-Navier and the Burnett approximations.

2) For the expansion coefficients α_{rl} of the distribution function, analogous theorems hold. At every stage of the approximation two further terms are obtained in the development in powers of μ, and these terms are unaltered by the higher approximations.

* The scheme for the higher approximation is that in the fourth and fifth approximation, one adds successively 4 rows and columns to the determinant; then in the next two orders of approximations one adds 5 rows and columns, and so on.

3) We know from sect. 3 that for $\sigma_0 = 0$ the equation has three zero roots and all other roots are pure imaginary. For $\sigma_0 \neq 0$ all these roots will become functions of σ_0, and may be called different modes of motion. Up to now we considered only the mode which in the zeroth order is given by $\omega_0^2 = 5\sigma_0^2/6$ and which therefore starts for $\sigma_0 = 0$ from two of the zero roots. For the mode starting from the third zero root one finds:

$$\omega^{(3)} = \frac{3}{2}\frac{i\mu\sigma^2}{\rho}\left\{1 - \frac{15}{4}\frac{\mu^2\sigma^2}{\rho^2 V_0^2}\cdots\right\} . \tag{32}$$

For real σ all terms are pure imaginary, so that this mode may be called *non-propagating*, in contrast to the mode starting from the other zero roots, given by eq. (29), which can also be written:

$$\omega^{(1,2)} = \pm V_0\sigma\left\{1 \pm \frac{7}{6}\frac{i\mu\sigma}{\rho V_0} + \frac{13}{8}\left(\frac{\mu\sigma}{\rho V_0}\right)^2 \pm \ldots\right\} . \tag{33}$$

Here all powers of σ occur in the series expansion and the terms are alternatingly real and imaginary. For real σ, ω is complex, so the mode is *propagating*. The question arises whether there are other propagating modes.

4) The form of the series expansions of the different modes in powers of σ follows from the fact * that Δ is an *even* function of σ. Since in any approximation Δ is a homogeneous polynomial in ω_0, σ_0, and the iA_k, and since the A_k are proportional to $1/\mu$, the dispersion equations has in any approximation the form:

$$P(x, y) = 0 , \tag{34}$$

where P is a polynomial in $x = i\mu\omega_0$ and $y = i\mu\sigma_0$ with real coefficients. From the theory of the algebraic functions it is known that all the branches starting from the distinct roots of eq. (34) in x for $y = 0$ must be analytic functions of y. Since only even powers of y can occur in eq. (34), it is seen that in these cases the form of the development is:

$$i\mu\omega_0 = a_0 + a_1(\mu\sigma_0)^2 + a_2(\mu\sigma_0)^4 \ldots ,$$

where the a_k's are real. Therefore all distinct roots will not give rise to propagating modes.

* This is physically obvious since σ and $-\sigma$ should always be equivalent; for a formal proof see appendix D.

5) Clearly the general situation will be that there are for $\sigma_0 = 0$ three zero roots and otherwise distinct and purely imaginary roots. One must expect, therefore, that in general there is only one propagating mode, corresponding to eq. (33). However, in the case of Maxwell molecules there are still "accidental" degeneracies. For instance, the eigenvalues λ_{11} and λ_{20} are equal and as a result the mode starting from these eigenvalues will have a form analogous to eq. (33). One finds in fact:

$$\omega = \frac{2i}{5}\frac{\rho V_0^2}{\mu}\left\{1 \pm \sqrt{5}\,\frac{\mu\sigma}{\rho V_0}\ldots\right\}.$$

This should therefore also be called a propagating mode, although since the first term is purely imaginary the damping will be much stronger, so that it is unlikely that these modes have any physical significance.

5. EXTENSION TO OTHER MOLECULAR MODELS; FINAL REMARKS [9*]

In principle, the method of sect. 3 can also be used for other force laws. However, there is then the practical difficulty that the eigenfunctions and eigenvalues of the collision operator are not known. In addition, the successive-approximation method described in sect. 4 is not so simple and straightforward.

The most obvious method for other molecular models is to use the *same* set of eigenfunctions in the expansion of the perturbation h as is used for Maxwell molecules *. Putting

$$h = \sum_{r'l'} \alpha_{r'l'}\psi_{r'l'}(c)$$

in eq. (15), multiplying by $\psi_{rl}\,\mathrm{e}^{-c^2}$, and integrating over $\mathrm{d}c$, one finds:

$$i\omega\alpha_{rl} - i\sqrt{\frac{2kT}{m}}\,\sigma\sum_{r'l'} M_{rl,r'l'}\alpha_{r'l'}$$

$$= n\sqrt{\frac{2kT}{m}}\sum_{r'l'} a_{rl,r'l'}\alpha_{r'l'}\,, \qquad (35)$$

where

* This is also essentially what is done in the Enskog-Chapman method.

$$a_{rl,r'l'} \equiv \int \mathrm{d}c \; e^{-c^2} \psi_{rl} \, J(\psi_{r'l'}) = -\pi^{\frac{3}{2}} \sqrt{\frac{m}{2kT}} \, [\psi_{rl}, \psi_{r'l'}] \, ,$$

using the bracket symbol of Chapman. The $a_{rl,r'l'}$ will be zero if $l' \neq l$ because of the isotropy of the collision operator J, but there will be no restriction on the "radial" numbers r and r'. If we now call:

$$\omega_1 = \frac{\omega}{n\sqrt{\frac{2kT}{m}}} \, , \qquad \sigma_1 = \frac{\sigma}{n} \, , \tag{37}$$

and arrange the determinant in the same manner as we did previously, then in the determinant Δ all element proportional to σ are unchanged. The diagonal elements will be of the form $\omega_1 + ia_{rl,rl}$. The only difference from the form of the determinant in sect. 3 is that there are new off-diagonal elements $ia_{rl,r'l}$.

Using the same successive-approximation method to derive the dispersion law for the propagating mode as before, one obtains in the zeroth approximation the same result as in the Maxwell case. This is, of course, to be expected, since the zeroth-order solution is governed solely by the conservation theorems. For the first approximation the expression for σ_1^2 is formally the same as eq. (27) in sect. 4, except that in place of λ_{02} and λ_{11} we have $a_{02,02}$ and $a_{11,11}$ respectively. A comparison with the Enskog-Chapman development reveals that $a_{02,02}$ and $a_{11,11}$ are related to μ_1 and ν_1, the "first approximation" of Chapman's expression for the viscosity and heat conduction coefficients, in exactly the same manner as the λ_{02} and λ_{11} are related to μ and ν in the Maxwell case. One finds, to the first order of small quantities:

$$\sigma_1^2 = \frac{6\omega_1^2}{5} \left\{ 1 + i\omega_1 \left[\frac{2}{5a_{11,11}} + \frac{4}{5a_{02,02}} \right] \cdots \right\} ,$$

which, when transformed to ordinary units, reads:

$$\sigma^2 = \frac{\omega^2}{V_0^2} \left\{ 1 - \frac{i\omega}{\rho V_0^2} \left(\frac{4}{15} \frac{m\nu_1}{k} + \frac{4}{3}\mu_1 \right) \cdots \right\} .$$

This is the same expression as that derived from the Stokes-Navier equations except that μ_1 and ν_1 take the place of μ and ν.

Going over to the second approximation, one finds, in contrast to the case of the Maxwell molecules, that the coefficient of the first approximation is changed. The result, to the first order of small quantities, is now

$$\sigma_1^2 = \frac{6\omega_1^2}{5}\left\{1 + i\omega_1\left[\frac{2}{5a_{11,11}} + \frac{4}{5a_{02,02}}\,\frac{a_{02,02}a_{12,12}}{a_{02,02}a_{12,12} - (a_{02,12})^2}\right]\cdots\right\}.$$

The change is that $1/a_{02,02}$ is replaced by:

$$\frac{1}{a_{02,02}}\,\frac{a_{02,02}a_{12,12}}{a_{02,02}a_{12,12} - (a_{02,12})^2},$$

which in Chapman's notation means that μ_1 is replaced by μ_2, the "second approximation" of the viscosity coefficient according to the method of Chapman.

The development proceeds in this manner. At every stage of the approximation the coefficients of *all* the lower-order terms change, and especially in the first-order term one gets successively higher approximations of the viscosity and heat conduction coefficients according to the method of Chapman.

This complication of our successive-approximation method for the dispersion law cannot be avoided, as one might perhaps surmise, by the use of the exact eigenfunctions of the collision operator, which would make the matrix $a_{rl,r'l'}$ diagonal in both r and l. Even with these eigenfunctions (supposing that one knew them!), one will in general not have the same selection rules for the "radial" number r in the matrix $M_{rl,r'l'}$ as one has for the case of the Maxwell molecules. As a result it is easily shown that in the second approximation the coefficient of the first-order or Stokes-Navier correction is modified.

From a practical point of view, this complication of the method is not very serious, since one can expect from the work of Chapman and Enskog that for the usual force laws the successive changes of the coefficients in each order will converge rapidly *. Clearly the origin of the method's complication lies in the fact that the arrangement of the determinant adopted for the Maxwell molecules is not suitable for other molecular models. We will now present an arrangement which will lead to "exact" expressions for the viscosity and heat conduction coefficients, and which presum-

* Compare the values of the successive approximations $\mu_1, \mu_2, \ldots$ and $\nu_1, \nu_2, \ldots$ of the viscosity and heat conduction coefficients in Chapman and Cowling, chap. 10.

ably will enable one to make a more consistent series-expansion development of the dispersion law. For this purpose we use the set of eigenfunctions belonging to the Maxwell model, so that the set of linear equations is the set given by eq. (25). In the arrangement of the determinant we leave the zeroth approximation unchanged but order the rest according to the "azimuthal number" l. In each block corresponding to a value of l, the arrangement is in the order of increasing values of r up to infinity. For the first approximation the blocks $l = 1$ and $l = 2$ are added to the zeroth-order determinant. The second approximation is obtained by the further addition of the blocks $l = 0$ and $l = 3$ and all values of r. For the mth approximation, where $m > 2$, one simply borders the determinant for the $(m - 1)$th approximation by the rows and columns corresponding to $l = m + 1$, and again all values of r. Using the notations defined by eqs. (36) and (37), the determinant for the first or the Stokes-Navier approximation is given below. The zeroth-order solution is

$$\sigma_1^2 = \frac{6\omega_1^2}{5},$$

while for the first approximation one easily obtains the equation:

$$\sigma_1^2 = \frac{6\omega_1^2}{5}\left\{1 + i\omega_1\left(\frac{2}{5}\frac{A_{11}}{A} + \frac{4}{5}\frac{B_{11}}{B}\right)\ldots\right\}, \tag{38}$$

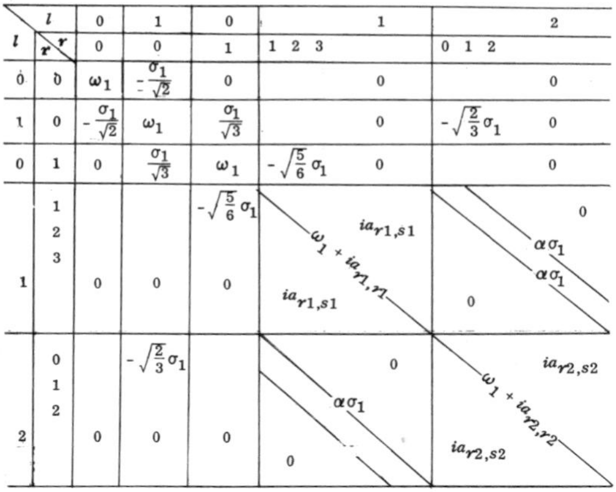

l	l	0	1	0	1	2
	r \ r	0	0	1	1 2 3	0 1 2
0	0	ω_1	$-\frac{\sigma_1}{\sqrt{2}}$	0	0	0
1	0	$-\frac{\sigma_1}{\sqrt{2}}$	ω_1	$\frac{\sigma_1}{\sqrt{3}}$	0	$-\sqrt{\frac{2}{3}}\sigma_1$ 0
0	1	0	$\frac{\sigma_1}{\sqrt{3}}$	ω_1	$-\sqrt{\frac{5}{6}}\sigma_1$ 0	0
1	1 2 3	0	0	$-\sqrt{\frac{5}{6}}\sigma_1$ 0	$ia_{r1,s1}$ $\omega_1 + ia_{r1,r1}$ $ia_{r1,s1}$	0 $\alpha\sigma_1$ $\alpha\sigma_1$ 0
2	0 1 2	0	$-\sqrt{\frac{2}{3}}\sigma_1$ 0	0	0 $\alpha\sigma_1$ 0	$ia_{r2,s2}$ $\omega_1 + ia_{r2,r2}$ $ia_{r2,s2}$

where $A = |a_{r1,s1}|$, r and s taking all values from 1 to ∞, and $B = |a_{r2,s2}|$ with r and s starting from 0. A_{11} and B_{11} are the first minors of A and B respectively when the corresponding first row and first column are struck out. The form of eq. (38) suggests the following identifications *:

$$\frac{\omega_1 A_{11}}{A} = -\frac{2}{3}\frac{m}{k}\frac{\nu\omega}{\rho V_0^2}, \tag{39a}$$

$$\frac{\omega_1 B_{11}}{B} = -\frac{5}{3}\frac{\mu\omega}{\rho V_0^2}, \tag{39b}$$

so that eq. (38) reduces in ordinary units to

$$\sigma^2 = \frac{\omega^2}{V_0^2}\left\{1 - \frac{i\omega}{\rho V_0^2}\left(\frac{4m}{15k}\nu + \frac{4}{3}\mu\right)\ldots\right\},$$

the same as the Stokes-Navier expression with μ and ν given by Chapman's formulas.

We conclude with some remarks on questions which have to be investigated further.

1) Clearly the outstanding problem is the question of the convergence of the series development for the dispersion law of the propagating mode, say, for Maxwell molecules. At every stage of approximation one gets a convergent series, as follows from the theory of algebraic functions. However, it is not sure that the regions of convergence in the successive stages of approximation have a region in common.

2) It would be very desirable to obtain an expression for the dispersion law which could be used for larger values of λ/Λ, so that a comparison with the experiments of Greenspan could be made. A development in powers of Λ/λ seems indicated, but we have not succeeded in obtaining such a series. It is perhaps possible to cast the infinite determinant into other analytical forms for the case of Maxwell molecules, and perhaps in this way a better discussion of the dispersion laws for the different modes can be made.

3) For a proper discussion of the experiments it is perhaps necessary to include the effect of the solid walls through which the sound vibrations are transmitted to the gas. In this connection it

* For justification of these identifications, see appendix E.

will be of interest to treat simple problems, like the Couette flow or the heat conduction between two parallel plates, where one has to take the effect of the walls into account. If the flow is slow, or the relative temperature difference small, then it is possible to formulate the problem for *all* values of λ/L (L = distance between plates) in a way analogous to the dispersion-of-sound problem.

4) These problems touch also the question of the calculation of the viscosity slip in gases as formulated recently by Kramers [10].

5) Finally, one should perhaps emphasize again that in the method proposed here it is in principle possible, for all types of flow problems, to separate sharply the effect of the (molecular) Mach number from the effect of the Knudsen number. For example, it will now perhaps be possible to discuss properly the drag of a sphere in a gas for small speeds, but for arbitrary ratio of the mean free path to the diameter of the sphere.

APPENDIX A. DEFINITION AND PROPERTIES OF $F(\theta)$

From the relationship between the angle θ through which the relative velocity is turned in a collision, and the collision parameter b, one obtains for Maxwell molecules:

$$\tfrac{1}{2}(\pi-\theta) = \int_0^{x'} \frac{dx}{\sqrt{1 - x^2 - \frac{1}{2}\left(\frac{x}{\alpha}\right)^4}} ,$$

where $x = b/r$, $\alpha = (mg^2/2\kappa)^{\frac{1}{4}}\, b$, and x' is the smallest positive root of the equation:

$$1 - x^2 - \frac{1}{2}\left(\frac{x}{\alpha}\right)^4 = 0 .$$

Making the substitution

$$\alpha^4 = 2 \cot^2 2\phi ,$$

so that the range $b = 0$ to $b = \infty$ corresponds to $\phi = \frac{1}{4}\pi$ to $\phi = 0$, one finds

$$\tfrac{1}{2}(\pi-\theta) = \sqrt{\cos 2\phi}\; K(\sin\phi) ,$$

where

$$K(\sin\phi) = \int_0^{\frac{1}{2}\pi} \frac{d\psi}{\sqrt{1 - \sin^2\phi \sin^2\psi}}$$

Table 3
Relationship between θ and ϕ for Maxwell molecules

ϕ	θ	ϕ	θ
0°0'	0° 0'	40° 0'	94°40'
5°0'	1° 2'	41° 0'	103° 4'
10°0'	4°10'	42° 0'	112°52'
15°0'	9°34'	43° 0'	124°44'
20°0'	17°30'	44° 0'	140°36'
25°0'	28°30'	44°30'	152° 2'
30°0'	43°24'	45° 0'	180° 0'
35°0'	63°58'		

is the complete elliptical integral of the first kind. This relationship between θ and ϕ has been tabulated by Maxwell [11] and is reproduced in table 3.

In terms of the variable ϕ, $F(\theta)$, which is defined as

$$F(\theta)\sin\theta\,\mathrm{d}\theta = \sqrt{\frac{m}{2\kappa}}\,gI(g,\theta)\sin\theta\,\mathrm{d}\theta = \sqrt{\frac{m}{2\kappa}}\,gb\,\mathrm{d}b\;,$$

is given by

$$F(\theta) = \frac{1}{2^{\frac{3}{2}}}\,\frac{\sqrt{\cos 2\phi}}{\sin\theta\,\sin 2\phi\,(\cos^2\phi\,K - \cos 2\phi\,E)}\;,$$

where

$$E = \int_0^{\frac{1}{2}\pi} \mathrm{d}\psi\,\sqrt{1-\sin^2\phi\,\sin^2\psi}$$

is the complete elliptical integral of the second kind. $F(\theta)$ is a monotonically decreasing function of θ. For small θ one has

$$F(\theta) \approx \frac{1}{8}\left(\frac{3\pi}{2}\right)^{\frac{1}{2}}\theta^{-\frac{5}{2}}\left(1+\frac{35}{24\pi}\,\theta+\ldots\right),$$

while for $\theta = \pi$ one has

$$F(\pi) = \frac{1}{4[K(\frac{1}{4}\pi)]^2} = 0.0727\;.$$

APPENDIX B. PROPERTIES OF THE SONINE POLYNOMIALS

The Sonine polynomial $S_n^{(m)}(x)$ is defined as the coefficient of s^m in the expansion of

$$(1-s)^{-n-1}\, e^{-xs/(1-s)} .$$

Its explicit expression is

$$S_n^{(m)}(x) = \sum_{p=0}^{m} (-x)^p \frac{(n+m)!}{p!\,(m-p)!\,(n+p)!} .$$

so that

$$S_n^{(0)} = 1 ,$$
$$S_n^{(1)} = n + 1 - x ,$$
$$S_n^{(2)} = \tfrac{1}{2}(n+1)(n+2) - (n+2)x + \tfrac{1}{2}x^2 ,$$

etc. The following integrals can be easily derived from the generating function:

1) Orthogonality relation

$$\int_0^\infty dx\, x^n\, e^{-x} S_n^{(m)} S_n^{(m')} = \frac{\Gamma(n+m+1)}{\Gamma(m+1)}\, \delta_{mm'}$$

2) $$\int_0^\infty dx\, x^n\, e^{-x} S_{n-1}^{(m)} S_n^{(m)} = \frac{\Gamma(n+m+1)}{\Gamma(m+1)} ,$$

3) $$\int_0^\infty dx\, x^n\, e^{-x} S_{n-1}^{(m+1)} S_n^{(m)} = -\frac{\Gamma(n+m+1)}{\Gamma(m+1)} .$$

APPENDIX C. THERE IS NO LOWER BOUND FOR THE SET OF EIGENVALUES

$$\lambda_{rl} = 2\pi \int_0^\pi d\theta \sin\theta\, F(\theta)\, \{\cos^{2r+l}(\tfrac{1}{2}\theta)\, P_l(\cos(\tfrac{1}{2}\theta))$$
$$+ \sin^{2r+l}(\tfrac{1}{2}\theta)\, P_l(\sin(\tfrac{1}{2}\theta)) - (1+\delta_{r0}\delta_{l0})\}.$$

We prove this statement in two steps:

1) $|\lambda_{rl}| > |\lambda_{r0}|$.

This is evident for $r = 0$. For $r \neq 0$

$$\lambda_{rl} - \lambda_{r0} = 2\pi \int_0^\pi d\theta \sin\theta\, F(\theta) \left\{\cos^{2r}(\tfrac{1}{2}\theta)\left[\cos^l(\tfrac{1}{2}\theta)\, P_l(\cos(\tfrac{1}{2}\theta)) - 1\right] + \sin^{2r}(\tfrac{1}{2}\theta)\left[\sin^l(\tfrac{1}{2}\theta))\, P_l(\sin(\tfrac{1}{2}\theta)) - 1\right]\right. .$$

Since $F(\theta)$ is always positive, and in the interval $0 \leqslant x \leqslant 1$

$$x^l P_l(x) \leqslant 1 \ ,$$

it follows that

$$\lambda_{rl} - \lambda_{r0} = \text{negative}$$

or

$$|\lambda_{rl}| > |\lambda_{r0}| \ ,$$

because λ_{rl} is negative. Hence if λ_{r0} has no lower bound, there will also be no lower bound for λ_{rl}.

2) There is no lower bound for λ_{r0}.
For $r \neq 0$,

$$\lambda_{r0} = 2\pi \int_0^\pi d\theta \sin\theta\, F(\theta)\, (\cos^{2r}(\tfrac{1}{2}\theta) + \sin^{2r}(\tfrac{1}{2}\theta) - 1) \ .$$

Since $F(\theta)$ diverges at $\theta = 0$ and decreases monotonically with increasing θ; and for large r, $(\cos^{2r}(\frac{1}{2}\theta) + \sin^{2r}(\frac{1}{2}\theta) - 1)$ is approximately -1 except for θ near 0 and π, we see that the main contribution to the integral will be from the neighborhood $\theta = 0$. Thus we can approximate $F(\theta)$ by

$$F(\theta) = \frac{1}{64}\sqrt{\frac{3\pi}{2}}\, \sin^{-\frac{5}{2}}(\tfrac{1}{2}\theta) \ ,$$

which has the correct behaviour for small θ, and lies always below the true $F(\theta)$ curve for $\theta > 0$. Further, we can drop the term $\sin^{2r}(\frac{1}{2}\theta)$, since its contribution goes to zero as $r \to \infty$. Changing the integration variable from θ to $x = \frac{1}{2}\theta$,

$$\lambda_{r0} \cong \frac{\pi}{8}\sqrt{\frac{3\pi}{2}} \int_0^{\frac{1}{2}\pi} dx \sin x \cos x\, \frac{1}{\sin^{\frac{5}{2}} x}\, (\cos^{2r} x - 1) = -\sqrt{\frac{3}{2}}\, \frac{\pi^{\frac{3}{2}}}{16} \sum_{i=0}^{r-1} \frac{\Gamma(i+1)\Gamma(\frac{3}{4})}{\Gamma(i+\frac{7}{4})} \ .$$

For large r, the ratio of the $i+1$ to the ith term of the sum is:

$$\frac{i+1}{i+\frac{7}{4}} \cong 1 - \frac{3}{4i}$$

Thus:

$$\lim_{r\to\infty} \lambda_{r0} = -\sqrt{\frac{3}{2}}\,\frac{\pi^{\frac{3}{2}}}{16}\,\Gamma(\tfrac{3}{4}) \lim_{r\to\infty} \sum_{i=0}^{r-1} \frac{\Gamma(i+1)}{\Gamma(i+\frac{7}{4})}$$

$$= -\infty .$$

APPENDIX D. PROOF THAT THE DETERMINANT Δ IS AN EVEN FUNCTION OF σ_0

This property of the determinant is independent of the interatomic forces. It is a consequence of the selection rule

$$l' = l \pm 1 .$$

We will assume h to be developed in the set of eigenfunctions ψ_{rl} of the linear operator J, and we will arrange the determinant according to the values of l. The determinant then takes the form:

	l'	0	1	2	3
l	r \ r'	0 1...	0 1...		
0	0 1 ⋮	×××××××× 0 0	$\propto \sigma_0$	0	0
1	0 1 ⋮	$\propto \sigma_0$	×××××××× 0 0	$\propto \sigma_0$	0
2		0	$\propto \sigma_0$	×××××××× 0 0	$\propto \sigma_0$
3		0	0	$\propto \sigma_0$	×××××××× 0 0

Along the main diagonal we have $\omega_0 + i\lambda_{rl}$, the non-diagonal elements in the diagonal blocks are all zero, and σ_0 appears only in the off-diagonal blocks as indicated. These blocks may be square or not, and they may have any number of rows and columns, depending on the atomic model and on the order of approximation. If we now change the signs of all the elements in the rows and the corresponding columns, all the signs of the σ_0's change to their negatives, while the signs along the main diagonal remain unchanged. The determinant Δ is, however, not changed because we have made an even number of sign changes and therefore we have shown that:

$$\Delta(\sigma_0) = \Delta(-\sigma_0) \; .$$

APPENDIX E. JUSTIFICATION FOR EQS. (39a) AND (39b)

By comparing the definitions of $a_{rl,sl}$ with the quantities a_{rs} and b_{rs} used by Chapman and Cowling [12], it is easily seen that:

$$a_{r1,s1} = -\frac{\pi^{\frac{3}{2}}}{3}\sqrt{\frac{m}{2kT}}\,N_{r1}N_{s1}\,a_{rs} \, ,$$

$$a_{r2,s2} = -\frac{3\pi^{\frac{3}{2}}}{10}\sqrt{\frac{m}{2kT}}\,N_{r2}N_{s2}\,b_{rs} \, ,$$

where N_{rl} is the normalization factor for ψ_{rl}, namely,

$$N_{rl} = \sqrt{\frac{r!\,(l+\frac{1}{2})}{\pi(l+r+\frac{1}{2})!}}$$

If we call $(A)_C = |a_{rs}|$, then:

$$A^{(p)} = (-)^p\left(\frac{\pi^{\frac{3}{2}}}{3}\right)^p\left(\frac{m}{2kT}\right)^{\frac{1}{2}p}(N_{11}N_{21}\cdots N_{p1})^2(A^{(p)})_C \, ,$$

$$A_{11}^{(p)} = (-)^{p-1}\cdot\left(\frac{\pi^{\frac{3}{2}}}{3}\right)^{p-1}\left(\frac{m}{2kT}\right)^{\frac{p-1}{2}}(N_{21}N_{31}\cdots N_{p1})^2(A_{11}^{(p)})_C \, ,$$

where the superscript p means that p rows and p columns are taken in A and $(A)_C$. Hence

$$\frac{\omega_1 A_{11}^{(p)}}{A^{(p)}} = -\frac{3}{n\pi^{\frac{3}{2}}}\,\frac{\omega}{N_{11}^2}\,\frac{(A_{11}^{(p)})_C}{(A^{(p)})_C} \; .$$

Since $N_{11} = \sqrt{\frac{4}{5}}\,\pi^{-\frac{3}{4}}$ and [12]

$$\lim_{p\to\infty} \frac{(A_{11}^{(p)})_C}{(A^{(p)})_C} = \frac{16}{225} \cdot \frac{3m\nu}{2k^2 T},$$

it follows that

$$\frac{\omega_1 A_{11}}{A} = \lim_{p\to\infty} \frac{\omega_1 A_{11}^{(p)}}{A} = -\frac{2}{3}\,\frac{m}{k}\,\frac{\nu\omega}{\rho V_0^2}.$$

Similarly, by making use of $N_{02} = \sqrt{\frac{4}{3}}\,\pi^{-\frac{3}{4}}$ and Chapman's result:

$$\lim_{p\to\infty} \frac{(B_{11}^{(p)})_C}{(B^{(p)})_C} = \frac{2\mu}{5kT},$$

one establishes eq. (39b).

REFERENCES

[1] H.Primakoff, J. Acous. Soc. Am. 14 (1942) 14.
[2] H.S.Tsien and R.Schamberg, J. Acous. Soc. Am. 18 (1946) 334.
[3] Chapman and Cowling, The mathematical theory of non-uniform gases, Chap. 15 (1st ed.; see note 2*).
[4] C.S.Wang Chang and G.E.Uhlenbeck, On the transport phenomena in rarified gases, Chap. I of this book.
[5] C.S.Wang Chang, On the dispersion of sound in helium, Chap. II of this book.
[6] M.Greenspan, Phys. Rev. 75 (1949) 197; J. Acous. Soc. Am. 22 (1951) 568.
[7] D.Enskog, Kinetische Theorie der Vorgange in Massig Verdunnten Gasen, Dissertation, Uppsala, 1917, p. 140.
[8] Watson, Bessel functions, p. 379.
[9] Watson, Bessel functions, eq. (3), p. 394.
[10] H.A.Kramers, Nuovo Cimento 6 (1949) 297.
[11] Maxwell, Collected papers II, p.42.
[12] Chapman and Cowling, The mathematical theory of non-uniform gases, Chap. 7.

Chapter V

THE KINETIC THEORY OF A GAS IN ALTERNATING OUTSIDE FORCE FIELDS: A GENERALIZATION OF THE RAYLEIGH PROBLEM

C. S. WANG CHANG and G. E. UHLENBECK
University of Michigan (1956)

1. STATEMENT OF THE PROBLEM

In this report we will be concerned with the following problem. Suppose a particle of mass m is bound harmonically to a fixed point with proper frequency ω_0; it is surrounded by a gas of particles of mass M against which it collides according to some given force law; the gas is supposed to be in equilibrium at temperature T and the equilibrium is *not* affected * by the motion of the particle m. Finally, an outside alternating force $mE_0 \cos \omega t$ acts, say in the x-direction, on the particle m (not on the molecules of the surrounding gas). One wants to know the average power absorbed by the particle as a function of ω_0, ω, the ratio of the masses m/M, and the type of force law between the particle and the molecules of the surrounding gas.

Clearly the problem is a generalization of the well-known problem of Rayleigh [1], in which it will go over if $\omega_0 = 0$, no outside force is present, and the motion is one-dimensional. The relation to the theory of the shape of absorption lines and to the theory of metals will be more or less evident and will not be further elaborated. The problem was in fact suggested in a discussion with Dr. J. M. Luttinger, because of a paradox which he encountered in the theory of metals.

* One may think that the gas is sufficiently dilute so that the velocity distribution of the molecules around the particle m remains the Maxwell distribution.

In the classical form in which we stated the problem, the mathematical formulation is given by the so-called *linear Boltzmann equation*. Let $f(\mathbf{x}, \mathbf{v}, t)\,\mathrm{d}\mathbf{x}\,\mathrm{d}\mathbf{v}$ be the probability at time t that the particle m is in the space and velocity range $\mathrm{d}\mathbf{x}\,\mathrm{d}\mathbf{v}$, then f will fulfill the equation

$$\frac{\partial f}{\partial t} + v_\alpha \frac{\partial f}{\partial x_\alpha} + a_\alpha \frac{\partial f}{\partial v_\alpha} = J(f) , \tag{1}$$

where $\boldsymbol{a}$ is the acceleration produced by the forces acting on the particle m, so that

$$\begin{aligned} a_x &= -\omega_0^2 x + E_0 \cos\omega t , \\ a_y &= -\omega_0^2 y , \qquad a_z = -\omega_0^2 z . \end{aligned} \tag{2}$$

$J(f)$ is the collision term:

$$J(f) = \int \mathrm{d}\mathbf{V} \int \mathrm{d}\Omega\, gI(g, \theta)\,[f'F' - fF] , \tag{3}$$

where

$$F(\mathbf{V}) = N\left(\frac{M}{2\pi kT}\right)^{\frac{3}{2}} \mathrm{e}^{-MV^2/2kT}$$

is the distribution function of the surrounding gas and the primes refer to the velocity variables; the collision $(\mathbf{v}, \mathbf{V}) \to (\mathbf{v}', \mathbf{V}')$ turns the relative velocity $g = |\mathbf{v} - \mathbf{V}|$ over the angle θ and $I(g, \theta)$ is the differential collision cross section.

The outside force $E_0 \cos\omega t$ must be considered as the perturbation which prevents the distribution function f from going to the equilibrium distribution:

$$f_0 = \left(\frac{m\omega_0}{2\pi kT}\right)^3 \mathrm{e}^{-(mv^2 + m\omega^2 r^2)/2kT} . \tag{4}$$

In the steady state we now want to calculate the time average of

$$P = \bar{v}_x E_0 \cos\omega t , \tag{5}$$

where

$$\bar{v}_x(t) = \iint \mathrm{d}\mathbf{x}\,\mathrm{d}\mathbf{v}\, v_x f(\mathbf{x}, \mathbf{v}, t) .$$

Clearly $f = f_0 + f_1$, where the perturbation f_1 of the distribution function will in the steady state be proportional to E_0 and vary in time like the outside force, although, of course, it will not be in phase because of the friction with the surrounding gas.

An exact solution of the problem we have found only for the case of the so-called Maxwell molecules, where one assumes that the interaction between particle m and a gas molecule is a repulsion $\propto 1/r^5$. Before turning to this special case we first will discuss some approximate solutions, assuming some approximate expressions for the collision operator which are current in the literature.

2. THE BROWNIAN MOTION LIMIT

If the particle m is very heavy compared to the gas molecule ($m/M \gg 1$), and if, in addition, we assume that the velocity $\boldsymbol{v}$ is never very different from the equipartition value, so that v/V is always of order $(M/m)^{\frac{1}{2}}$, then one finds. by an expansion in powers of M/m, that the collision term $J(f)$ can be approximated by the well-known Rayleigh or Brownian motion form

$$J(f) \approx \eta \left[\frac{\partial}{\partial v_\alpha} (v_\alpha f) + \frac{kT}{m} \frac{\partial^2 f}{\partial v_\alpha \partial v_\alpha}\right], \tag{6}$$

where the friction coefficient η is given by *

$$\eta = \frac{16\sqrt{\pi}}{3} \frac{NM}{m} \left(\frac{M}{2kT}\right)^{\frac{5}{2}} \int_0^\pi d\theta \sin\theta (1-\cos\theta) \int_0^\infty dV V^5 e^{-MV^2/2kT} I(V,\theta).$$

Since for general $I(g,\theta)$ the proof of eq. (6) is not easily available, we give the details in appendix A.

With the collision term (6) it is simple to solve the problem. Multiplying the Boltzmann equation

$$\frac{\partial f}{\partial t} = -v_\alpha \frac{\partial f}{\partial x_\alpha} - a_\alpha \frac{\partial f}{\partial v_\alpha} + \eta \frac{\partial}{\partial v_\alpha} \left[v_\alpha f + \frac{kT}{m} \frac{\partial f}{\partial v_\alpha}\right] \tag{7}$$

with x_i or v_i, respectively, and integrating over the coordinate and velocity space, assuming that f vanishes sufficiently fast for large x_i and v_i, one obtains for the average values $\overline{x}_i$ and $\overline{v}_i$ the equations

* The assumption that v/V is always of order $(M/m)^{\frac{1}{2}}$ implies that on the average the particle m will feel a frictional force proportional to its velocity; the proportionality constant is η. For very large v, this will not be true anymore; the friction will then become "Newtonian", proportional to v^2.

$$\frac{d\bar{x}_i}{dt} = \bar{v}_i , \qquad \frac{d\bar{v}_i}{dt} = -\omega_o^2 \bar{x}_i + E_o \cos \omega t \, \delta_{i1} - \eta \bar{v}_i , \tag{8}$$

which have an obvious physical interpretation. In the steady state

$$\bar{v}_i = \frac{E_o \eta \omega^2}{(\omega^2 - \omega_o^2)^2 + \eta^2 \omega^2} \cos \omega t + \frac{E_o \omega (\omega^2 - \omega_o^2)}{(\omega^2 - \omega_o^2)^2 + \eta^2 \omega^2} \sin \omega t ,$$

so that the average power absorbed is given by

$$\bar{P}_{BM} = \tfrac{1}{2} E_o^2 \eta \frac{\omega^2}{(\omega^2 - \omega_o^2)^2 + \eta^2 \omega^2} .$$

For a discussion of this result, see sect. 4.

3. THE STRONG COUPLING APPROXIMATION

Especially in the theory of metals, it is customary to approximate the collision term by assuming

$$J(f) \approx \frac{f_o - f}{\tau} , \tag{10}$$

where f_o is the equilibrium distribution (4) and τ is the *relaxation time*, which is a measure of the time required for the collisions to establish equilibrium. One of the main questions we will have to discuss is the question under which circumstances (10) may be used as an approximation of the collision term.

With eq. (10), one obtains for the equation of motion of the average values $\bar{x}_i, \bar{v}_i$ instead of eq. (8), the equations

$$\frac{d\bar{x}_i}{dt} = \bar{v}_i - \frac{\bar{x}_i}{\tau} , \qquad \frac{d\bar{v}_i}{dt} = -\omega_o^2 \bar{x}_i + E_o \cos \omega t \, \delta_{i1} - \frac{\bar{v}_i}{\tau} . \tag{11}$$

Note especially the first of these two equations. It says that the average position of the probability distribution does *not* change with time according to the average velocity $\bar{v}_i$. The origin of this paradoxical result is the fact that with (10)

$$\int d\mathbf{v}\, J(f) = \frac{1}{\tau} \int d\mathbf{v}\, (f_o - f) ,$$

which is not necessarily zero, while from the exact expression (3) follows

$$\int \mathrm{d}\boldsymbol{v}\, J(f) = 0 \ . \tag{12}$$

Eq. (12) is an expression of the fact that in a collision the number of particles does not change. One must say therefore that the strong collision approximation (10) violates this conservation law. A consequence of this is, as Luttinger has pointed out, that it makes a difference whether one calculates the average power absorbed with the help of the average velocity or with the help of $\mathrm{d}\bar{x}_i/\mathrm{d}t$. Using the average velocity, one obtains from eq. (11) the result first derived by Van Vleck and Weisskopf [2]:

$$\bar{P}_{\mathrm{VW}} = \frac{E_0^2\tau}{4}\left[\frac{1}{1+(\omega-\omega_0)^2\tau^2} + \frac{1}{1+(\omega+\omega_0)^2\tau^2}\right], \tag{13}$$

while, using $\mathrm{d}\bar{x}/\mathrm{d}t$, one obtains,

$$\bar{P}_{\mathrm{L}} = \frac{E_0^2\tau}{4}\,\frac{\omega}{\omega_0}\left[\frac{1}{1+(\omega-\omega_0)^2\tau^2} - \frac{1}{1+(\omega+\omega_0)^2\tau^2}\right], \tag{14}$$

first given by Luttinger [3].

4. DISCUSSION OF THE APPROXIMATE RESULTS

If one puts in the Brownian motion result (9) $\eta = 1/\tau$, then the three results, (9), (13), and (14), can be directly compared with each other. In figs. 1 to 5 we have plotted $\bar{P}$ as a function of $\omega\tau$ for various values of $\omega_0\tau$. One easily verifies the following facts:

a) For all three forms, the area under the curve is the same and equal to $\frac{1}{4}\pi E_0^2$.

b) For small $\omega_0\tau$, $\bar{P}_{\mathrm{VW}}$ will be a monotonic decreasing function of $\omega\tau$. Only for $\omega_0\tau > 1/\sqrt{3}$, $\bar{P}_{\mathrm{VW}}$ will have a maximum. Since both $\bar{P}_{\mathrm{L}}$ and $\bar{P}_{\mathrm{BM}}$ are zero for $\omega\tau = 0$, if $\omega_0\tau \neq 0$, they always will show a maximum.

c) For $\omega_0\tau = 0$, $\bar{P}_{\mathrm{VW}}$ and $\bar{P}_{\mathrm{BM}}$ are identical.

d) For $\omega_0\tau \gg 1$, $\bar{P}_{\mathrm{VW}}$ and $\bar{P}_{\mathrm{L}}$ become nearly the same, especially near the resonance peak. The $\bar{P}_{\mathrm{BM}}$ gives an essentially sharper resonance peak.

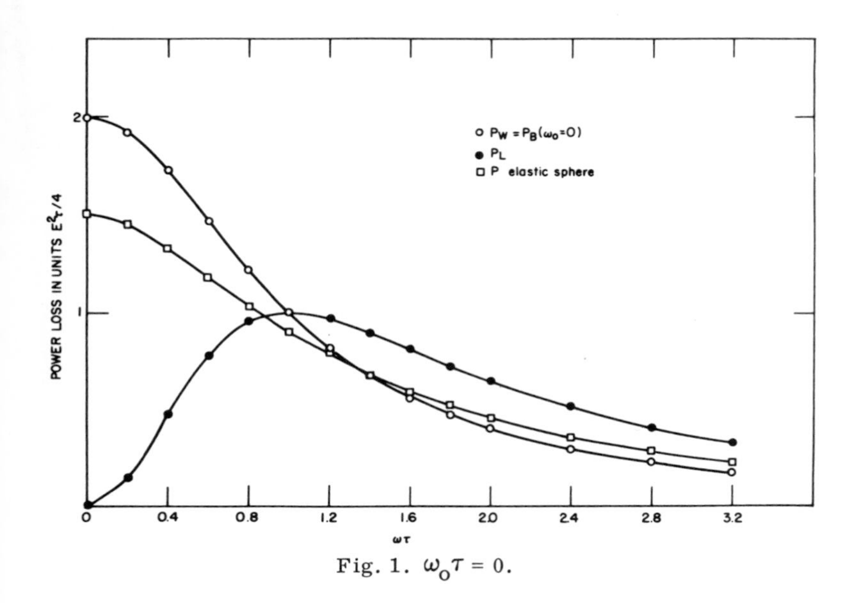

Fig. 1. $\omega_0\tau = 0$.

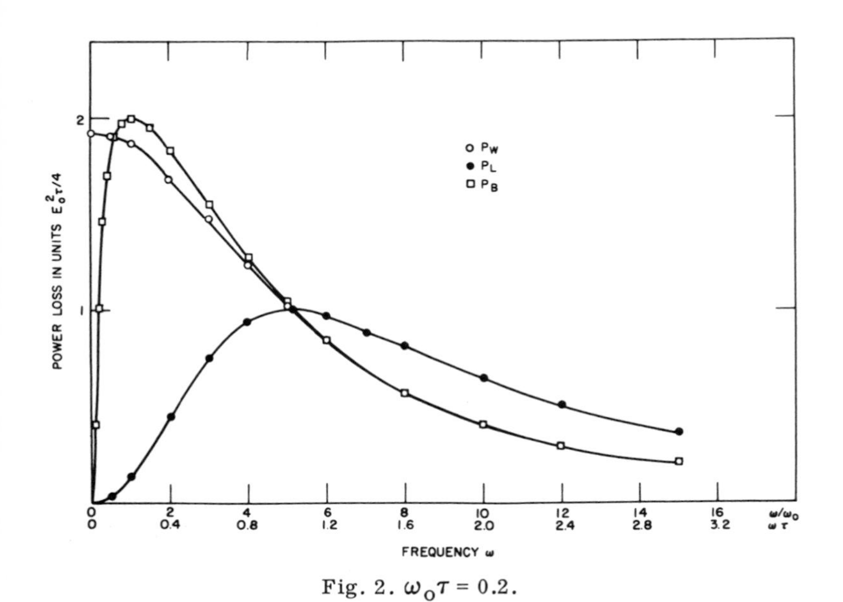

Fig. 2. $\omega_0\tau = 0.2$.

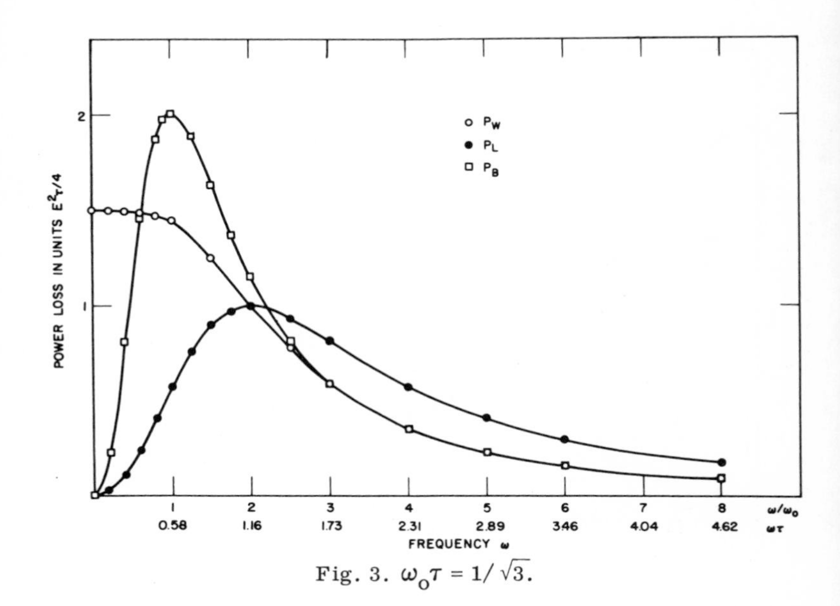

Fig. 3. $\omega_0\tau = 1/\sqrt{3}$.

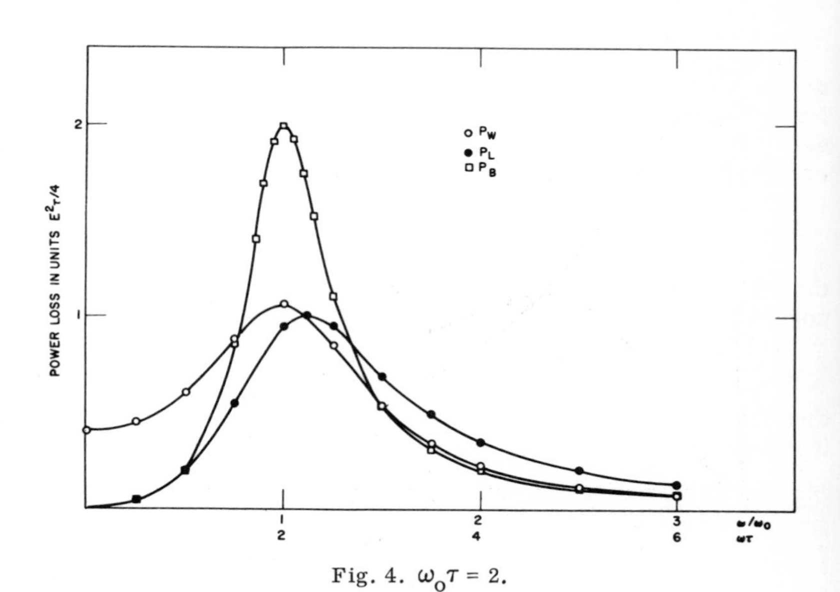

Fig. 4. $\omega_0\tau = 2$.

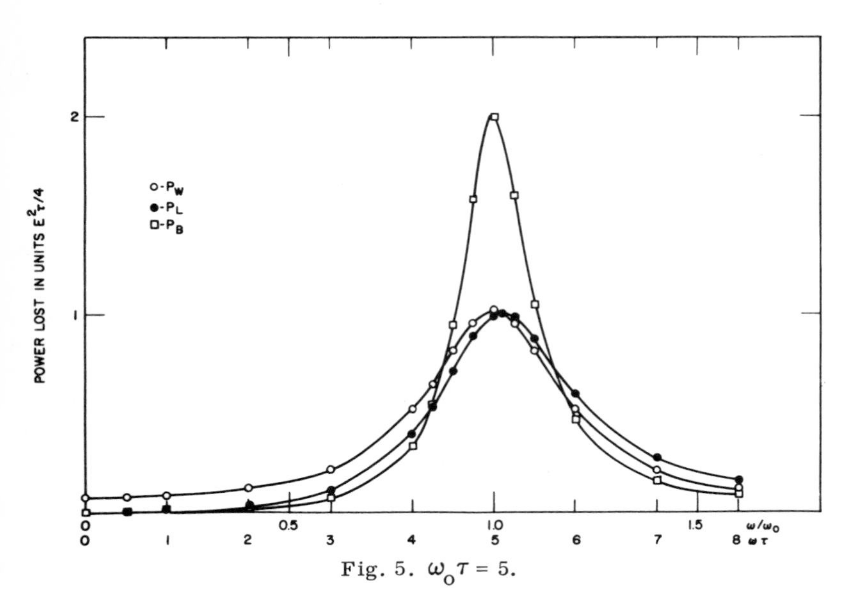

Fig. 5. $\omega_0\tau = 5$.

5. THE EXACT SOLUTION FOR MAXWELL MOLECULES

Although not quite necessary for the solution of our problem, it seems worthwhile first to point out that for Maxwell molecules and for arbitrary ratio of the masses, the eigenfunctions and eigenvalues of the collision operator can be determined. For Maxwell molecules $gI(g,\theta)$ is independent of g. Putting

$$gI(g,\theta) = \sqrt{\frac{2K(M+m)}{Mm}}\, F(\theta)\ , \tag{15}$$

then $F(\theta)$ is the dimensionless function discussed in a previous report [4]. Introducing in the exact collision term (3)

$$f = e^{-mv^2/2kT}\, h\ ,$$

then from the energy conservation in the collision $(v, V) \rightarrow (v', V')$ it follows that

$$J(f) = N\sqrt{\frac{2K(M+m)}{Mm}}\, e^{-mv^2/2kT}\, I(h)\ , \tag{17}$$

where $I(h)$ is the dimensionless collision operator

$$I(h) = \frac{1}{\pi^{\frac{3}{2}}} \int \mathrm{d}\boldsymbol{C}\ \mathrm{e}^{-C^2} \int \mathrm{d}\Omega\ F(\theta)\,(h' - h) \tag{18}$$

and $\boldsymbol{C} = \boldsymbol{V}(M/2kT)^{\frac{1}{2}}$. It is also convenient to consider h as a function of the dimensionless velocity $\boldsymbol{c} = \boldsymbol{v}\ (m/2kT)^{\frac{1}{2}}$ and then h' means $h(\boldsymbol{c}')$, where

$$\boldsymbol{c}' = \boldsymbol{c} + \frac{2M}{m+M}\ \boldsymbol{s}\,[\boldsymbol{s}\cdot(\sqrt{m/M}\,\boldsymbol{C} - \boldsymbol{c})] \tag{19}$$

is the dimensionless velocity of the particle after collision *.

Since the operator I does not depend on the velocity $\boldsymbol{c}$, and since the connection between $\boldsymbol{c}$ and $\boldsymbol{c}'$ is linear, it is clear that the eigenfunctions of I must be polynomials in $\boldsymbol{c}$. Since the operator I is spherically symmetric in the velocity space $\boldsymbol{c}$, the dependence of the eigenfunctions on the direction of $\boldsymbol{c}$ must be like a spherical harmonic. In fact, one can prove that the eigenfunctions are of the form

$$\psi_{rlm}(\boldsymbol{c}) = N_{rlm} c^l\, Y_{lm}(\theta,\phi)\, S^{(r)}_{l+\frac{1}{2}}(c^2)\ , \tag{20}$$

where

$$S^{(r)}_{l+\frac{1}{2}}(c^2)$$

is the Sonine polynomial of degree r and order $l + \frac{1}{2}$ and N_{rlm} is a normalization factor. The ψ_{rlm} form a complete orthogonal set of functions with the weight factor $\exp(-c^2)$. The corresponding eigenvalues are given by

$$\lambda_{rl} = 2\pi \int_0^{\pi} \mathrm{d}\theta\ \sin\theta\ F(\theta) \left\{ \left[1 - \frac{4mM}{(m+M)^2} \sin^2(\tfrac{1}{2}\theta) \right]^{r+\frac{1}{2}l} P_l(\cos\psi) - 1 \right\}. \tag{21}$$

P_l is the Legendre polynomial, and

$$\cos\psi = \left[1 - \frac{4mM}{(m+M)^2} \sin^2(\tfrac{1}{2}\theta) \right]^{-\frac{1}{2}} \left(1 - \frac{2M}{m+M} \sin^2(\tfrac{1}{2}\theta) \right). \tag{21a}$$

For a proof of these statements see appendix B. Note that the eigenfunctions (20) are *independent* of the mass ratio m/M, which enters only in the eigenvalues (21). The first eigenvalue $\lambda_{00} = 0$, corresponding to $\psi_{000}(\boldsymbol{c})$ = constant; this expresses the conservation of the number of particles in a collision. All other eigenvalues

* The symbol $\boldsymbol{s}$ is the unit vector in the direction of closest approach; $\boldsymbol{s}$ makes an angle $\frac{1}{2}(\pi+\theta)$ with $\boldsymbol{g} = \boldsymbol{C} - \boldsymbol{c}$.

are negative, and this expresses the tendency of f to go to the Maxwell distribution.

The reason why our problem of the power loss can be solved exactly for Maxwell molecules lies in the fact that the velocity c is an eigenfunction * of the collision operator I, corresponding to $r = 0$, and $l = 1$. By developing h in the eigenfunctions ψ_{rlm}, one sees that because of the orthogonality property of the ψ_{rlm}

$$\int \mathrm{d}c\ c_i\ \mathrm{e}^{-c^2}\ I(h) = \lambda_{01} \int \mathrm{d}c\ c_i\ \mathrm{e}^{-c^2} h\ . \tag{22}$$

As a consequence, one obtains from the Boltzmann equation for the average values $\bar{x}_i$ and $\bar{v}_i$ the equations

$$\frac{\mathrm{d}\bar{x}_i}{\mathrm{d}t} = \bar{v}_i\ , \qquad \frac{\mathrm{d}\bar{v}_i}{\mathrm{d}t} = -\omega_0^2 x_i + E_0 \cos\omega t\ \delta_{i1} - \eta\bar{v}_i\ , \tag{23}$$

where η is given by

$$\eta = -N\sqrt{\frac{2K(M+m)}{Mm}}\ \lambda_{01}$$
$$= 2\pi N\sqrt{\frac{2K(M+m)}{Mm}}\ \frac{M}{m+M}\int_0^\pi \mathrm{d}\theta\ \sin\theta(1-\cos\theta)F(\theta)\ .$$

We see therefore that for all ratios of the masses one obtains the *same* functional dependence as in the Brownian motion limit. In the limit $m/M \gg 1$, the value of η goes over into the value found in the Brownian motion limit (see appendix A). In addition, one can show as a check that the eigenfunctions of the Brownian motion form (6) of the collision operator are again the ψ_{rlm} given by eq. (20), while the eigenvalues are equidistant and equal to

$$\Lambda_{rl} = -\eta\,(2r+l)\ .$$

For the proof, see appendix C.

It is therefore clear that the strong collision approximation discussed in sect. 3 cannot have a general validity independent of the intermolecular forces. Especially, it cannot be true that in the limit $m/M \ll 1$, which is opposite to the Brownian motion limit ‡, the collision term can be approximated by the strong coupling form (10) for all types of intermolecular force laws. Of course, it may be that the inverse fifth power law gives too "soft" collisions.

* Note that $S_k^{(0)}(x) = 1$.

‡ It may be called the Lorentz limit, since it corresponds to the situation considered in the Lorentz theory of electronic conduction in metals.

It is therefore of interest to investigate other force laws and especially the case of elastic spheres.

6. THE SOLUTION FOR ELASTIC SPHERES IN THE LORENTZ LIMIT IF $\omega_0 = 0$

If $m/M \ll 1$ and $\omega_0 = 0$, then our problem can be solved by an adaptation of the perturbation method used in the Lorentz theory of electronic conduction in metals. The distribution function f now depends only on the velocity $\mathbf{v}$ and the time and fulfills the equation

$$\frac{\partial f}{\partial t} + E_0 \, e^{i\omega t} \frac{\partial f}{\partial v_x} = J(f) \; . \tag{24}$$

Considering the second term as the perturbation which prevents f from reaching the equilibrium distribution

$$f_0 = \left(\frac{m}{2\pi kT}\right)^{\frac{3}{2}} \exp\left(-mv^2/2kT\right) \; ,$$

one obtains, by putting

$$f = f_0(1+h) \; , \tag{25}$$

for h the inhomogeneous equation

$$\frac{\partial h}{\partial t} - \frac{mv_x}{kT} E_0 \, e^{i\omega t} = \int \mathrm{d}\mathbf{V}\, F(V) \int \mathrm{d}\Omega \, gI(g,\theta)(h' - h) \; . \tag{26}$$

Because of the linearity, h will be $\propto \exp(i\omega t)$ in the steady state, and in the limit $m/M \ll 1$, h will have the form

$$h = h_0(v^2) v_x \, e^{i\omega t} \; , \tag{27}$$

since in the lowest order of m/M the velocity v does not change in magnitude in a collision, and g may be replaced by v. Hence, substituting eq. (27) in the right-hand side of eq. (26), one gets

$$\int \mathrm{d}\mathbf{V}\, F(V) \int \mathrm{d}\Omega \, vI(v,\theta) h_0(v^2)(v_x' - v_x)$$
$$= -2\pi N h_0(v^2) v_x \int_0^{\pi} \mathrm{d}\theta \, \sin\theta \, (1-\cos\theta) \, vI(v,\theta) \; .$$

Therefore, for elastic spheres $[I(v,\theta) = \frac{1}{4}\sigma^2$, where σ is the average of the diameters of the spheres with masses m and $M]$ one obtains

$$h_0(v^2) = \frac{mE_0}{kT} \frac{1}{i\omega + \pi\sigma^2 Nv} . \tag{28}$$

Introducing the relaxation time

$$\tau = \frac{1}{\pi\sigma^2 N} \sqrt{\frac{m}{2kT}} , \tag{29}$$

one gets for the average velocity in the x-direction

$$\bar{v}_x = \frac{8E_0\tau}{3\sqrt{\pi}} e^{i\omega t} \int_0^\infty dc\, c^4 e^{-c^2} \frac{1}{c + i\omega\tau} , \tag{30}$$

and hence for the average power absorbed

$$\bar{P} = \frac{4E_0^2\tau}{3\sqrt{\pi}} \int_0^\infty dc\, c^5 e^{-c^2} \frac{1}{c^2 + \omega^2\tau^2}$$

$$= \frac{2E_0^2\tau}{3\sqrt{\pi}} [1 - \omega^2\tau^2 + \omega^4\tau^4 e^{+\omega^2\tau^2} \mathrm{Ei}(-\omega^2\tau^2)] , \tag{31}$$

where $\mathrm{Ei}(y)$ is the exponential integral

$$\mathrm{Ei}(-y) = \int_y^\infty dx \frac{e^{-x}}{x} .$$

$\bar{P}$ as function of ω is shown in fig. 1. One easily verifies that the area under the curve is again $\frac{1}{4}\pi E_0^2$, just as for all the other curves. One sees that the elastic-sphere result lies between the Van Vleck-Weisskopf and the Luttinger results. In fact, for large $\omega\tau$

$$\bar{P}_{\mathrm{VW}} \propto \frac{2}{\omega^2\tau^2} , \qquad \bar{P}_{\mathrm{L}} \propto \frac{4}{\omega^2\tau^2} ,$$

while from eq. (31) one obtains

$$\bar{P} \propto \frac{16}{3\sqrt{\pi}\omega^2\tau^2} = \frac{3.01}{\omega^2\tau^2} .$$

One can solve the problem also in another way, which is of interest since it may be generalizable to the case where ω_0 is not zero. Expand the perturbation h in the eigenfunctions (20) of the collision operator for Maxwell molecules. Then we can write

$$f = f_0 \left[1 + \sum_{r,l} \alpha_{rl} \psi_{rl}(c)\, e^{i\omega t} \right] , \tag{32}$$

where $c = v(m/2kT)^{\frac{1}{2}}$ and α_{rl} are the development coefficients *. Substituting in the Boltzmann equation (24), multiplying by $\psi_{r'l'} \times \exp(-c^2)$, and integrating over $\boldsymbol{c}$, one obtains a set of linear equations in α_{rl}:

$$i\omega\alpha_{rl} - \sqrt{\frac{m}{2kT}}\frac{2E_0}{N_{01}}\delta_{r0}\delta_{l1} = \sum_{r',l'}[\psi_{rl}, \psi_{r'l'}]\alpha_{r'l'} , \tag{33}$$

where the bracket symbols are defined by:

$$[\psi_{rl}, \psi_{r'l'}] = \int \mathrm{d}\boldsymbol{c}\ \mathrm{e}^{-c^2}\psi_{rl}(\boldsymbol{c}) \int \mathrm{d}\boldsymbol{V} F(V) \times \int \mathrm{d}\Omega\, gI(g,\theta)[\psi_{r'l'}(\boldsymbol{c}') - \psi_{r'l'}(\boldsymbol{c})] .$$

The N_{rl} are normalization constants, determined by

$$\int \mathrm{d}\boldsymbol{c}\ \exp(-c^2)\,\psi_{rl}^2 = 1 ,$$

which gives

$$N_{rl} = \sqrt{\frac{r!\,(l+\frac{1}{2})}{\pi(l+\frac{1}{2}+r)!}} .$$

For arbitrary ratio of m/M the calculation of the bracket symbol is complicated, but in the Lorentz limit $m/M \ll 1$ the result is again simple. As shown in appendix D one obtains in this limit for elastic spheres:

$$[\psi_{r'l'}, \psi_{rl}] = -\frac{1}{2\tau}\frac{N_{rl}N_{r'l'}}{2l+1}\delta_{ll'}\sum_s \frac{(s+l+1)!\,(r'-\frac{3}{2}-s)!\,(r-\frac{3}{2}-s)!}{s!\,(r'-s)!\,(r-s)!} . \tag{34}$$

Since the equations (33) are not coupled through the index l, and since the inhomogeneous part contains only $l = 1$, we can restrict ourselves throughout to $l = 1$. The equations (33) can then be written in the form

$$i\omega\alpha_r - \frac{1}{\tau}\sum_{r'=0}^{\infty} b_{rr'}\alpha_{r'} = \sqrt{\frac{m}{2kT}}\frac{2E_0}{N_{01}}\delta_{r0} , \tag{35}$$

with $r = 0, 1, 2, \ldots$ and where $b_{rr'}$ is the bracket symbol $\tau[\psi_{r1}, \psi_{r'1}]$. The average velocity in the x-direction depends on α_0; in fact,

$$\bar{v}_x = \tfrac{1}{2}N_{01}\alpha_0 \sqrt{\frac{2kT}{m}}\,\mathrm{e}^{i\omega t} .$$

* Because of the axial symmetry of the problem we can take $m = 0$.

From eq. (35) follows

$$\alpha_0 = \sqrt{\frac{m}{2kT}}\,\frac{2E_0}{N_{01}}\,\tau\,\frac{D_{00}(i\omega\tau)}{D(i\omega\tau)}\,,$$

where D is the determinant $||i\omega\tau\,\delta_{rr'} - b_{rr'}||$ and D_{00} is the minor of the (0, 0) element. Thus

$$\bar{v}_x = E_0\tau\,\frac{D_{00}(i\omega\tau)}{D(i\omega\tau)}\,e^{i\omega\tau}\,, \tag{36}$$

which must be compared with eq. (30). The identity of eqs. (30) and (36) for $\omega = 0$ has been shown by Chapman [5]. We also verified the identity for large ω. A complete formal proof of the identity is lacking.

7. CONCLUDING REMARKS

If ω_0 is not zero, then the perturbation h will depend on the coordinates as well as on $\boldsymbol{v}$ and t. It seems feasible to generalize the second method of the previous section by developing h into products $\psi_{rlm}(\boldsymbol{x})\psi_{rlm}(\boldsymbol{v})$, using the same type of functions in both $\boldsymbol{x}$ and $\boldsymbol{v}$, and again considering the Lorentz limit. However, the details have not yet been worked out.

Presumably for elastic spheres and in the Lorentz limit, the result for $\omega_0 \neq 0$ will always lie between the Van Vleck-Weisskopf and the Luttinger result. There is one feature which the exact result will have in common with the Luttinger result, namely, that for $\omega_0 \neq 0$ and $\omega = 0$, $\bar{P}$ will be zero. This is clear, because in this case the constant outside force will only polarize the oscillator and in the steady state f will be the Maxwell-Boltzmann distribution

$$f \propto \exp\left(-\frac{m}{2kT}\right)\left\{v^2 + \omega_0^2\left[\left(x + \frac{E_0}{\omega_0^2}\right)^2 + y^2 + z^2\right]\right\}\,,$$

so that $\bar{v}_x$ is zero.

APPENDIX A. PROOF OF THE BROWNIAN MOTION FORM OF THE COLLISION OPERATOR

As in sect. 5 we put in the exact collision term (3)

$$f = e^{-mv^2/2kT}\, h\,, \tag{1.A}$$

then again it follows that

$$J(f) = N\left(\frac{M}{2\pi kT}\right)^{\frac{3}{2}} e^{-mv^2/2kT}\, I(h)\,, \tag{2.A}$$

where

$$I(h) = \int \mathrm{d}\boldsymbol{V}\; e^{-MV^2/2kT} \iint \mathrm{d}\epsilon\, \mathrm{d}\theta\, \sin\theta\; gI(g,\theta)(h' - h)\,. \tag{3.A}$$

Since for $m/M \gg 1$, $\boldsymbol{v}'$ differs very little from v, one may make a Taylor expansion:

$$h' - h = (v'_\alpha - v_\alpha)\frac{\partial h}{\partial v_\alpha} + \frac{1}{2}(v'_\alpha - v_\alpha)(v'_\beta - v_\beta)\frac{\partial^2 h}{\partial v_\alpha \partial v_\beta} + \ldots\,. \tag{4.A}$$

From the momentum conservation follows

$$\boldsymbol{v}' - \boldsymbol{v} = -\frac{M}{m+M}(\boldsymbol{g}' - \boldsymbol{g})\,, \tag{5.A}$$

where $\boldsymbol{g} = \boldsymbol{V} - \boldsymbol{v}$, $\boldsymbol{g}' = \boldsymbol{V}' - \boldsymbol{v}'$, are the relative velocities before and after collision *. Introducing (4.A) and (5.A) one can integrate over the azimuthal angle ϵ, since only $\boldsymbol{g}' - \boldsymbol{g}$ depends on ϵ. Next, introduce in the velocity space $\boldsymbol{V}$ polar coordinates with the direction of $\boldsymbol{v}$ as polar axis. One then can again integrate over the azimuthal angle of $\boldsymbol{V}$. In these two integrations over azimuthal angles, the following general formulas are used, which are easily verified:

$$\int_0^{2\pi} \mathrm{d}\Phi\,(A_i - B_i) = 2\pi\left(\frac{A}{B}\cos\Theta - 1\right) B_i$$

$$\int_0^{2\pi} \mathrm{d}\Phi\,(A_i - B_i)(A_j - B_j)$$

$$= 2\pi\left\{B_i B_j\left[\left(\frac{A}{B}\cos\Theta - 1\right)^2 - \frac{1}{2}\frac{A^2}{B^2}\sin^2\Theta\right] + \frac{1}{2}A^2\sin^2\Theta\,\delta_{ij}\right\}.$$

In here $\boldsymbol{B}$ is a fixed vector and Θ, Φ are the polar angles of the vector $\boldsymbol{A}$ with respect to $\boldsymbol{B}$.

Now it is convenient to use dimensionless variables:

$$\boldsymbol{c} = \sqrt{\frac{m}{2kT}}\,\boldsymbol{v}\,, \qquad \boldsymbol{C} = \sqrt{\frac{M}{2kT}}\,\boldsymbol{V}. \tag{6.A}$$

* Of course, $|\boldsymbol{g}'| = |\boldsymbol{g}|$, and θ is the angle between $\boldsymbol{g}'$ and $\boldsymbol{g}$.

Note that we use different units for v and V. Assuming, as we will from now on, that c and C are of the same order of magnitude means that we have introduced our second assumption namely, that v never differs very much from the equipartition value. With these units

$$g = \sqrt{\frac{2kT}{M}}\, C \left(1 - 2\sqrt{\frac{M}{m}}\frac{c}{C}\cos\phi + \frac{M}{m}\frac{c^2}{C^2}\right)^{\frac{1}{2}},$$

where ϕ is the angle between C and c. Up to order $(m/M)^{\frac{1}{2}}$, one has therefore

$$gI(g,\theta) = \sqrt{\frac{2kT}{M}}\, CI\left(\sqrt{\frac{2kT}{M}}\, C,\theta\right)\left[1 - \sqrt{\frac{M}{m}}\frac{c}{C}\cos\phi\left(1 + \frac{C}{c}\frac{\partial I}{\partial C}\right)\right].$$

Developing also the rest of the integrand in powers of M/m, one obtains

$$I(h) = 4\pi^2 \frac{M}{m}\left(\frac{2kT}{M}\right)^2 \int_0^\pi d\theta \sin\theta\,(1-\cos\theta) \int_0^\infty dC\, C^3\, e^{-C^2}$$

$$\times \int_0^\pi d\phi \sin\phi\; I\left(\sqrt{\frac{2kT}{M}}\, C,\theta\right)\left[1 - \sqrt{\frac{M}{m}}\frac{c}{C}\cos\phi\left(1 + \frac{C}{I}\frac{\partial I}{\partial C}\right)\right]$$

$$\times \left\{\sqrt{\frac{m}{M}}\cos\phi\,\frac{C}{c}\, c_\alpha \frac{\partial h}{\partial c_\alpha} - \left[c_\alpha \frac{\partial h}{\partial c_\alpha} + \frac{1}{8}(3\cos\theta - 1)(3\cos^2\phi - 1)\right.\right.$$

$$\left.\times \frac{C^2}{c^2}\, c_\alpha c_\beta \frac{\partial^2 h}{\partial c_\alpha \partial c_\beta} + \frac{1}{8}(\cos^2\phi - 3 - \cos\theta\,(3\cos\phi - 1))\, C^2 \frac{\partial^2 h}{\partial c_\alpha \partial c_\alpha}\right]$$

$$\left. + \, O\left(\sqrt{M/m}\right)\right\}.$$

Carrying out the ϕ integration, keeping under the integral sign only the terms of order one*, and making the partial integration

$$\int_0^\infty dC\, C^4\, e^{-C^2} \frac{\partial I}{\partial C} = -\int_0^\infty dC\, e^{-C^2}(4C^3 - 2C^5)\, I\,,$$

one obtains

* It needs some further argument to show that it would be inconsistent to include higher-order terms in M/m, since then also the further terms in the Taylor development (4.A) would have to be included.

$$I(h) = \frac{8\pi^2}{3}\,\frac{M}{m}\left(\frac{2kT}{M}\right)^2 \int_0^\pi d\theta\,\sin\theta\,(1-\cos\theta)\int_0^\infty dC\,e^{-C^2}$$
$$\times C^5 I\left(\sqrt{\frac{2kT}{M}}\,C,\theta\right)\left(\frac{\partial^2 h}{\partial c_\alpha \partial c_\alpha} - 2c_\alpha \frac{\partial h}{\partial c_\alpha}\right) .$$

Introducing this expression in eq. (2.A), and going back to the original velocity variables $\boldsymbol{v}$ and $\boldsymbol{V}$, and to the original distribution function f, one gets

$$J(f) = \eta \frac{\partial}{\partial v_\alpha}\left(v_\alpha f + \frac{kT}{m}\frac{\partial f}{\partial v_\alpha}\right) ,$$

with

$$\eta = \frac{16\sqrt{\pi}}{3} N \frac{M}{m}\left(\frac{M}{2kT}\right)^{\frac{5}{2}} \int_0^\pi d\theta\,\sin\theta\,(1-\cos\theta)$$
$$\times \int_0^\infty dV\,V^5\,e^{-MV^2/2kT} I(V,\theta) .$$

For a repulsive force Kr^{-s}, we write for the differential cross section

$$I(g,\theta) = g^{-4/(s-1)} F(\theta, K, s) ,$$

and then

$$\eta = \frac{8\sqrt{\pi}}{3}\,\frac{NM}{m}\left(\frac{2kT}{M}\right)^{(s-5)/2(s-1)} \Gamma\left(\frac{3s-5}{s-1}\right) \int_0^\pi d\theta\,\sin\theta\,(1-\cos\theta) F(\theta,K,s).$$

Especially, for Maxwell molecules ($s = 5$) one has

$$\eta = 2\pi N \frac{M}{m}\sqrt{\frac{2kT}{M}} \int_0^\pi d\theta\,\sin\theta\,(1-\cos\theta) F(\theta) ,$$

where $F(\theta) = [mM/2K(m+M)]^{\frac{1}{2}} F(\theta,K,5) \approx (M/2K)^{\frac{1}{2}} F(\theta,K,5)$ is the dimensionless collision cross section used in sect. 5, and for elastic spheres ($s = \infty$, σ = diameter) *,

$$\eta = \frac{8\sqrt{\pi}}{3}\,\frac{NM\sigma^2}{m}\sqrt{\frac{2kT}{M}} .$$

* For elastic spheres the Brownian motion form of the collision operator was first derived by M.S.Green, J. Chem. Phys. 19 (1951) 1036.

APPENDIX B. EIGENVALUES AND EIGENFUNCTIONS OF THE MAXWELL COLLISION OPERATOR

The dimensionless collision operator

$$I(h) = \frac{1}{\pi^{\frac{3}{2}}} \int \mathrm{d}\boldsymbol{c}\; \mathrm{e}^{-c^2} \int \mathrm{d}\Omega\, F(\theta)\, [h(\boldsymbol{c}') - h(\boldsymbol{c})]$$

can be written in the standard form [6] (note that A_0 is divergent and really should be kept together with the second term)

$$I(h) = -A_0 h + \frac{1}{\pi^{\frac{3}{2}}} \left(\frac{M}{m}\right)^{\frac{3}{2}} \mathrm{e}^{c^2} \int \mathrm{d}\boldsymbol{c}'\, K(\boldsymbol{c}, \boldsymbol{c}')\, h(\boldsymbol{c}') ,$$

where

$$A_0 = 2\pi \int_0^{\pi} \mathrm{d}\theta \, \sin\theta \, F(\theta) ,$$

$$K(\boldsymbol{c}, \boldsymbol{c}') = 4\pi \left(\frac{m+M}{M}\right)^3 \exp\left\{-\frac{m+M}{2m}\left[c^2 + c'^2 + \frac{m-M}{2M}(\boldsymbol{c}-\boldsymbol{c}')^2\right]\right\}$$

$$\times \int_0^{\pi} \mathrm{d}\theta\, \cos(\tfrac{1}{2}\theta)\, \csc^2(\tfrac{1}{2}\theta)\, F(\theta) \exp\left[-\frac{(m+M)^2}{4mM}(\boldsymbol{c}-\boldsymbol{c}')^2 \cot^2(\tfrac{1}{2}\theta)\right]$$

$$\times J_0\left(-i\,\frac{m+M}{m}\, cc' \sin(\boldsymbol{c}, \boldsymbol{c}')\cot(\tfrac{1}{2}\theta)\right) .$$

We now will verify that

$$\psi_{rlm} = c^l S^{(r)}_{l+\frac{1}{2}}(c^2)\, Y_{lm}(\phi\chi)$$

is an eigenfunction of I. Let ϕ', χ' be the polar angles of $\boldsymbol{c}'$ with respect to the same set of axes as used for $\boldsymbol{c}$. For the integration over $\boldsymbol{c}'$ we take the direction of $\boldsymbol{c}$ as polar axis and let ϕ_1, χ_1 then be the polar angles. The integration over χ_1 can be performed, using

$$\int_0^{2\pi} \mathrm{d}\chi_1\, Y_{lm}(\phi', \chi') = 2\pi\, P_l(\cos\phi_1)\, Y_{lm}(\phi, \chi) , \tag{7.A}$$

and one obtains

$$I(\psi_{rlm}) = -A_o \psi_{rlm} + \sqrt{\pi}\left(\frac{M+m}{\sqrt{mM}}\right)^3 e^{-(m-M)^2 c^2/4mM}$$

$$\times Y_{lm}(\phi,\chi) \int_0^\pi d\theta \cos(\tfrac{1}{2}\theta)\csc^2(\tfrac{1}{2}\theta) F(\theta)\, e^{-\frac{(M+m)^2}{4mM} c^2 \cot^2(\frac{1}{2}\theta)}$$

$$\times \int_0^\pi dc'\, c'^{l+2} S^{(r)}_{l+\frac{1}{2}}(c'^2)\, e^{-\frac{(m+M)^2}{4mM} c'^2 \csc^2(\frac{1}{2}\theta)}$$

$$\times \int_0^\pi d\phi_1 \sin\phi_1 P_l(\cos\phi_1) J_o\left(-i\frac{m+M}{m} cc' \sin\phi_1 \cot(\tfrac{1}{2}\theta)\right)$$

$$\times \exp\ 2cc'\left[\frac{m^2-M^2}{4mM} + \frac{(m+M)^2}{4mM}\cot^2(\tfrac{1}{2}\theta)\right]\cos\phi_1 \quad .$$

Putting

$$z = -i\,\frac{(m+M)^2}{2mM}\csc^2(\tfrac{1}{2}\theta)\sqrt{1-\frac{4mM}{(m+M)^2}\sin^2(\tfrac{1}{2}\theta)}\; cc' \equiv \alpha c' \ ,$$

$$\cos\psi = \frac{1-\dfrac{2M}{m+M}\sin^2(\tfrac{1}{2}\theta)}{\sqrt{1-\dfrac{4mM}{(m+M)^2}\sin^2(\tfrac{1}{2}\theta)}} \ ,$$

then the last integral can be carried out, using

$$\int_0^\pi d\phi_1 \sin\phi_1 P_l(\cos\phi_1)\, e^{iz\cos\phi_1\cos\psi} J_o(z\sin\phi_1\sin\psi)$$

$$= \sqrt{\frac{2\pi}{z}}\, i^l P_l(\cos\psi) J_{l+\frac{1}{2}}(z) \ .$$

Writing $z = \alpha c'$, the integral over c' can be carried out next with the help of the formula

$$\int_0^\infty dc'\, c'^{l+\frac{3}{2}} S^{(r)}_{l+\frac{1}{2}}(c'^2)\, e^{-q^2c'^2} J_{l+\frac{1}{2}}(\alpha c')$$

$$= \frac{\alpha^{l+\frac{1}{2}}}{(2q^2)^{l+\frac{3}{2}}}\left(\frac{q^2-1}{q}\right)^r e^{-\alpha^2/4q^2} S^{(r)}_{l+\frac{1}{2}}\left[-\frac{\alpha^2}{4q^2(q^2-1)}\right] .$$

where q^2 is an abbreviation for

$$q^2 = \frac{(m+M)^2}{4mM} \csc^2 (\tfrac{1}{2}\theta) .$$

Putting everything together, one obtains

$$I(\psi_{rlm}) = -A_0 \psi_{rlm} + 2\pi \psi_{rlm}$$

$$\times \int_0^{\pi} \mathrm{d}\theta \sin\theta\, F(\theta)\, P_l(\cos\psi) \left[1 - \frac{4mM}{(m+M)^2} \sin^2(\tfrac{1}{2}\theta)\right]^{r+\frac{1}{2}l} ,$$

which is $\lambda_{rl}\psi_{rlm}$, where the eigenvalue λ_{rl} is given by the equations (21) and (21a).

APPENDIX C. EIGENVALUES AND EIGENFUNCTIONS OF THE BROWNIAN MOTION FORM OF THE COLLISION OPERATOR

The question is to find the eigenfunctions and eigenvalues of the differential equation

$$\frac{\partial}{\partial v_\alpha}\left(v_\alpha f + \frac{kT}{m}\frac{\partial f}{\partial v_\alpha}\right) = \frac{\Lambda}{\eta} f .$$

Introducing the dimensionless velocity $c = v\,(m/2kT)^{\frac{1}{2}}$, we find

$$\Delta f + c_\alpha \frac{\partial f}{\partial c_\alpha} + 6f = \frac{2\Lambda}{\eta} f .$$

The angular dependence is clearly like a spherical harmonic, and putting

$$f = c^l R(c)\, \mathrm{e}^{-c^2} Y_{lm}(\theta, \phi) ,$$

one gets for the radial function R,

$$\frac{\mathrm{d}^2R}{\mathrm{d}c^2} + \left[\frac{2(l+1)}{c} - 2c\right]\frac{\mathrm{d}R}{\mathrm{d}c} - \left[2l + \frac{2\Lambda}{\eta}\right] R = 0 ,$$

or using $x = c^2$ as independent variable,

$$x\frac{\mathrm{d}^2R}{\mathrm{d}x^2} + (l + \tfrac{3}{2} - x)\frac{\mathrm{d}R}{\mathrm{d}x} - \left(\frac{l}{2} + \frac{2\Lambda}{\eta}\right) R = 0 .$$

Comparing this equation with the equation for the Sonine polynomial $S_t^{(r)}(x)$,

$$x\frac{\mathrm{d}^2 S_t^{(r)}}{\mathrm{d}x^2} + (t+1-x)\frac{\mathrm{d}S_t^{(r)}}{\mathrm{d}x} + r S_t^{(r)} = 0 ,$$

one sees that $R = S^{(r)}_{l+\frac{1}{2}}(c\ )$ and that $r = -(\frac{1}{2}l + 2\Lambda/\eta)$, so that the eigenvalue

$$\Lambda_{lr} = -\eta(2r + l) .$$

It is also easy to verify that the exact formula (21) in sect. 5 leads in the Brownian motion limit to the same result.

APPENDIX D. THE BRACKET EXPRESSION $[\psi_{r'l'm'}, \psi_{rlm}]$ IN THE LORENTZ LIMIT

In the limit $m/M \ll 1$ we will calculate the bracket expression for a repulsive intermolecular force equal to Kr^{-s}, since this contains the Maxwell model and the elastic-sphere model as limiting cases. Writing as in appendix A,

$$gI(g, \theta) = g^{(s-5)/(s-1)} F(\theta, K, s) ,$$

the bracket expression becomes

$$[\psi_{r'l'm'}, \psi_{rlm}] = N\left(\frac{m}{2kT}\right)^{\frac{3}{2}} \left(\frac{M}{2\pi kT}\right)^{\frac{3}{2}} \int d\boldsymbol{v}\ e^{-mv^2/2kT}$$

$$\times \int d\boldsymbol{V}\ e^{-MV^2/2kT} \int_0^{\pi} \int_0^{2\pi} d\epsilon\, d\theta \sin\theta\, g^{(s-5)/(s-1)} F(\theta, K, s)$$

$$\times\, \psi_{r'l'm'}\left(\boldsymbol{v}\sqrt{\frac{m}{2kT}}\right)\left[\psi_{rlm}\left(\boldsymbol{v}'\sqrt{\frac{m}{2kT}}\right) - \psi\left(\boldsymbol{v}\sqrt{\frac{m}{2kT}}\right)\right] .$$

Since mv^2 is of the same order of magnitude as MV^2, in the limit $m/M \ll 1$ and to the lowest order in m/M, one can replace $\boldsymbol{g}'$ by $-\boldsymbol{v}'$ and $\boldsymbol{g}$ by $-\boldsymbol{v}$. Using eq. (7.A) of appendix B, the integral over ϵ can be carried out, and also the integral over $\boldsymbol{V}$ is immediate. With dimensionless velocity variables, one then obtains

$$[\psi_{r'l'm'}, \psi_{rlm}] = 2\pi N\left(\frac{2kT}{m}\right)^{(s-5)/2(s-1)} \int_0^{\pi} d\theta \sin\theta\, F(\theta, K, s)$$

$$\times [P_l(\cos\theta) - 1] \int d\boldsymbol{c}\ e^{-c^2} c^{(s-5)/(s-1)}\, \psi_{r'l'm'}(\boldsymbol{c})\, \psi_{rlm}(\boldsymbol{c}) .$$

Since the angular dependence of the ψ_{rlm} is a spherical harmonic, the integral over the directions of $\boldsymbol{c}$ is immediate. Putting in the explicit expression for the Sonine polynomials, the integral over c

can be carried out in each term and the result is a double sum, of which one sum can be evaluated with the help of the formula

$$\sum_s \binom{m}{k-s} \binom{n+s}{n} (-1)^s = \binom{m-n-1}{k} .$$

One obtains

$$[\psi_{r'l'm'}, \psi_{rlm}] = \frac{4\pi^2}{2l+1} N_{rl} N_{r'l'} \delta_{ll'} \delta_{mm'} \left(\frac{2kT}{m}\right)^{(s-5)/2(s-1)}$$

$$\times \int_0^\pi d\theta \sin\theta \, F(\theta, K, s) [P_l(\cos\theta) - 1] \frac{1}{\left\{\left[-1 - \frac{s-5}{2(s-1)}\right]!\right\}^2}$$

$$\times \sum_p \frac{\left[p + \frac{1}{2}(l+l') + \frac{1}{2} + \frac{s-5}{2(s-1)}\right]! \left[r - p - 1 - \frac{s-5}{2(s-1)}\right]! \left[r' - p - 1 - \frac{s-5}{2(s-1)}\right]!}{p!(r'-p)!(r-p)!}$$

Since

$$\frac{1}{2} \geqslant \frac{s-5}{2(s-1)} \geqslant 0 \qquad \text{for} \qquad \infty \geqslant s \geqslant 5 ,$$

only the case of Maxwell molecules ($s = 5$) needs special consideration since then factorials of negative integers appear. In this case one finds either directly or by a limit consideration,

$$[\psi_{r'l'm'}, \psi_{rlm}] = \frac{4\pi^2 N}{2l+1} N_{rl} N_{r'l'} \delta_{rr'} \delta_{ll'} \delta_{mm'}$$

$$\times \frac{(r+l+\frac{1}{2})!}{r!} \int_0^\pi d\theta \sin\theta \, F(\theta, K, 5) [P_l(\cos\theta) - 1] .$$

Finally, for elastic spheres ($s = \infty$), one gets

$$[\psi_{r'l'm'}, \psi_{rlm}] = \frac{8\pi\sigma^2 N}{2l+1} N_{rl} N_{r'l'} \delta_{ll'} \delta_{mm'} \sqrt{\frac{2kT}{m}}$$

$$\times \sum_p \frac{(p+l+1)!(r-\frac{3}{2}-p)!(r'-\frac{3}{2}-p)!}{p!(r'-p)!(r-p)!}$$

which reduces to the expression (34) used in sect. 6.

REFERENCES

[1] Rayleigh, Scientific papers, Vol. 3, p. 473;
Ming Chen Wang, A study of various solutions of the Boltzmann equation, Dissertation, Univ. of Michigan, Ann Arbor, 1942.
[2] J.H. Van Vleck and V.F. Weisskopf, Rev. Mod. Phys. 17 (1945) 227.
[3] J.M. Luttinger, private communication.
[4] C.S. Wang Chang and G.E. Uhlenbeck, On the propagation of sound in monoatomic gases, Chap. IV of this book.
[5] S. Chapman, J. London Math. Soc. 8 (1933) 266.
[6] D. Enskog, Kinetische Theorie der Vorgänge in mässig verdünnten Gasen, Dissertation, Uppsala, 1917, p. 154.

NOTES

[1*] After the work of J. R. Dorfman, E. G. D. Cohen, K. Kawasaki and I. Oppenheim we now know that this statement is *not* correct. The dependence of the transport coefficients on the pressure or density is *not* similar to the virial expansion of the equation of state, since terms containing ln p appear. For a recent review see the article of E. G. D. Cohen on the kinetic theory of dense gases in Fundamental Problems in Statistical Mechanics, Vol. II (North-Holland Publ. Comp., Amsterdam, 1968).

[2*] These errors appeared in the first edition of the book of Chapman and Cowling. They have been corrected in the second edition quoted in [1].

[3*] This work has been incorporated in the joint article with J. de Boer which appeared in Studies in Statistical Mechanics, Vol. II Part C (North-Holland) Publ. Comp., Amsterdam, 1964). There one finds also more recent literature.

[4*] The macroscopic causality problem has been clarified very much by the work of Bogoliubov, which was reprinted in Vol. I of the Studies in Statistical Mechanics. For a recent discussion see the article by one of us (G. E. U.) in Fundamental Problems in Statistical Mechanics, Vol. II, p. 19.

[5*] The special interest of this report is the result that the dispersion of sound for Maxwellian molecules given in Chapter I, Sect. 5, is almost the same as for the more realistic Lennard Jones potential if one expresses $(\lambda/\Lambda)^2$ always in terms of the viscosity coefficient. This insensitivity with regard to the intermolecular force law was completely confirmed by J. Foch (Dissertation, Rockefeller University, 1967) and this makes a significant comparison with experiment possible. See J. Foch and G. E. Uhlenbeck, Phys. Rev. Letters 19 (1967) 1025, and the complete discussion by J. Foch and G. W. Ford in Vol. VI of the Studies.

[6*] As mentioned in the foreword these results of Mrs. Chang have so far not been compared with experiment. An analysis of the deviations from the Navier-Stokes result, similar to the analysis used in the dispersion of sound, would be of great interest.

[7*] This publication was unfortunately never completed.

[8*] The eigenfunctions and eigenvalues of the linearized collision operator for Maxwell molecules were found by Mrs. Chang and later independently by L. Waldmann (Handbuch der Physik, Vol. 12 §38). For a very complete tabulation see Z. Alterman, K. Frankowski and C. L. Pekeris, Astrophys. J. Suppl. Series 7 (1962) 291.

[9*] A much better method for deriving the dispersion law for other molecular models was developed by J. Foch, see note [5*]. For hard spheres compare also the paper by C. L. Pekeris and co-workers, Phys. of Fluids 5 (1962) 1608.

Part B

The dispersion of sound in monoatomic gases

JAMES D. FOCH Jr.
Bell Telephone Laboratories
(Present address: University of Colorado)

and

GEORGE W. FORD
University of Michigan

Chapter I

INTRODUCTION

It is generally believed that the Boltzmann equation is the last word in the description of a dilute monatomic gas. Despite its widespread acceptance, however, the equation has not been checked beyond the Navier-Stokes level of hydrodynamics (where there is good agreement between the calculated and observed temperature dependence for the coefficients of viscosity and thermal conductivity). The question thus arises: Are there additional consequences of the Boltzmann equation which can be used as a more exacting test of its validity?

The purpose of this study is to give an account of attempts to check the Boltzmann equation through the investigation of sound propagation, and to indicate the considerable work which remains to be done. There are several reasons why sound propagation has gradually been singled out [1]. First, the Navier-Stokes and other higher order hydrodynamic equations (derived from the Boltzmann equation) can be applied to sound propagation without considering the vexing problem of boundary conditions. Second, the cumbersome derivation of hydrodynamic equations from the Boltzmann equation can be avoided completely by basing the theory of sound propagation directly on the linearized Boltzmann equation. Third, available experiments show clearly that the Navier-Stokes order of hydrodynamic equations cannot account for the observed phenomena quantitatively.

Before we proceed to the "dynamics" of sound propagation, we wish to make a few remarks about the corresponding kinematics. In general sound is a small amplitude disturbance of the gas which may be considered as a superposition of plane waves. For each of these the pressure variation Δp is of the form

$$\Delta p = A\, e^{i(\boldsymbol{k}\cdot\boldsymbol{r}-\omega t)} , \qquad (1.1)$$

in which A is a complex amplitude, ω is the circular frequency of the wave, and $\boldsymbol{k}$ is the propagation vector. If we take the direction of $\boldsymbol{k}$ to be the z axis then (1.1) becomes

$$\Delta p = e^{i(kz - \omega t)} \quad . \tag{1.2}$$

If the disturbance is maintained by the continuous operation of a source of fixed frequency then for each z the time dependence of Δp will be sinusoidal for all t, and hence ω must be considered real [2]. The sign of ω is positive for a wave emitted along the positive z axis and negative for one along the negative z axis; $|\omega|$ is just the source frequency. In this case the magnitude of the propagation vector must be complex, in order that there be no disturbance sufficiently far from the source.

Thus, still for a continuous source, one can write

$$k = \frac{2\pi}{\lambda} + i\alpha \tag{1.3}$$

with λ the sound wavelength and α the reciprocal of the absorption length (the distance in which the amplitude decreases by e^{-1}). The phase velocity of the wave is

$$v_{\mathrm{ph}} = \frac{\lambda}{2\pi}\,\omega = \left[\mathrm{Re}\left(\frac{k}{\omega}\right)\right]^{-1} \quad . \tag{1.4}$$

In general the phase velocity and absorption length are functions of ω. In fact, α must be an odd function of ω if the disturbance is to vanish far from the source along *both* the positive and negative z axes. Similarly, v_{ph} must be an even function of ω if there is no anisotropy to distinguish between sound propagation along positive and negative z axes.

On the other hand, if the disturbance of the gas is not maintained but rather evolves from some initial value, then $k = 2\pi/\lambda$, where λ is the wavelength of the initial disturbance, and k is real. The frequency ω must now be complex in order that there be no disturbance after a sufficiently long time. Thus, we can write

$$\omega = \omega' - i\frac{1}{\tau} \ , \tag{1.5}$$

where τ is the relaxation time (the time in which the amplitude decreases by e^{-1}). The phase velocity of the wave is now given by

$$v_{\mathrm{ph}} = \frac{\lambda}{2\pi}\,\omega' = \mathrm{Re}\left(\frac{\omega}{k}\right) , \tag{1.6}$$

which in general is not simply related to (1.4). Of course, if the damping of the waves is small, then (1.4) and (1.6) are equal for corresponding waves and

$$\alpha^{-1} = v_{\text{ph}} \tau \ . \tag{1.7}$$

This discussion of kinematics shows that whether one choses ω or k real in the plane waves such as (1.1) depends upon whether one is describing the source problem or the initial value problem. In any event, what we shall soon see is that the result of theoretical analysis is always a dispersion law, i.e. a complex function $F(\omega, k)$ which when set equal to zero,

$$F(\omega, k) = 0 \ , \tag{1.7}$$

gives the relation between ω and k. For the source problem one solves for k as a function of real ω, while for the initial value problem one solves for ω as a function of real k. The dispersion law, however, is always the same.

The first theory of sound propagation in gases was given by Newton. In 1687 he published (in the *Principia*) the erroneous result that the speed of sound in a gas is given by the formula

$$v = \left\{ \frac{p}{\rho} \right\}^{\frac{1}{2}} \ , \tag{1.8}$$

where p is the pressure and ρ the mass density of the gas. The origin of the error is that Newton assumed the compressions and rarefractions in a sound wave are isothermal, while in fact they are adiabatic. This error was corrected by Laplace, who showed that the correct formula is

$$v = \left\{ \left(\frac{\partial p}{\partial \rho} \right)_{\text{ad}} \right\}^{\frac{1}{2}} \ , \tag{1.9}$$

which relates the sound speed to the adiabatic compressibility. For an ideal gas

$$\left(\frac{\partial p}{\partial \rho} \right)_{\text{ad}} = \frac{c_p}{c_v} \frac{p}{\rho} \equiv \gamma \frac{p}{\rho} \ . \tag{1.10}$$

where c_p and c_v are the specific heat at constant pressure and volume, respectively. Note the absence of frequency dependence in (1.9); i.e. there is no dispersion. For monatomic gases, where $\gamma \approx 1.67$, Laplace's expression is about 30% greater than Newton's. Until the latter part of the nineteenth century the only real justification for preferring the Laplace result over that of Newton was that it fits the experiments much better. Not until the work of

Stokes and Kirchhoff was the Laplace result put on a more firm foundation and corrections to it arising from the failure of the adiabatic assumption determined.

Following the historical development, we begin our account in chapter II by considering the problem from the point of view of the hydrodynamic equations, which is really the method of Kirchhoff. As the sequel will show, however, it is both feasible and profitable to turn directly to the Boltzmann equation and to study sound dispersion on that basis. This we do in chapters III and IV. In chapter V we discuss model equations which resemble the Boltzmann equation and which are amenable to exact analysis. We conclude in chapter VI with an outlook.

NOTES

1) Experiments on light scattering by dilute monatomic gases, although at present much less accurate than sound propagation experiments, will also lead to tests of the Boltzmann equation [see e.g. M. Nelkin and S. Yip, Phys. Fluids 9 (1966) 380]. However, for a theory of light scattering one needs to know the time correlation of spontaneous density fluctuations, which is related to the initial value problem for sound propagation through the fluctuation-dissipation theorem [for a discussion of this theorem, see L. Onsager and S. L. Machlap, Phys. Rev. 91 (1953) 1505]. But the fluctuation-dissipation theorem, like the Boltzmann equation, is a theorem of macroscopic physics, which does not have a completely satisfactory derivation based on the microscopic laws of molecular motion. Thus, in light scattering the interpretation of experiment rests on the fluctuation-dissipation theorem as well as the Boltzmann equation, which one would like to test separately.

2) The most extensive experiments on sound propagation in dilute monatomic gases are still those of M. Greenspan, J. Acous. Soc. Amer. 28 (1956) 644, and they are of the source problem type. [See also E. Meyer and G. Sessler, Z. Physik 149 (1957) 15]. The sound is generated by a piezoelectric transmitter and detected by a movable piezoelectric receiver. As the distance L between transmitter and receiver is changed, the phase and amplitude of the signal from the receiver change, and from this one can infer the phase velocity and absorption length.

Chapter II

SOUND DISPERSION FROM THE HYDRODYNAMIC EQUATIONS

1. THE LINEARIZED HYDRODYNAMIC EQUATIONS

In this chapter we shall study the dispersion of sound by means of the continuum equations of hydrodynamics. This is the usual method which is known, at least in an elementary form, to every physics student. Our starting point is the general hydrodynamic equations [1]:

$$\frac{\partial \rho}{\partial t} + \frac{\partial \rho u_\alpha}{\partial x_\alpha} = 0$$

$$\rho \frac{\mathrm{D} u_i}{\mathrm{D}t} = - \frac{\partial P_{i\alpha}}{\partial x_\alpha} ,$$

$$\rho \frac{\mathrm{De}}{\mathrm{D}t} = - \frac{\partial q_\alpha}{\partial x_\alpha} - P_{\alpha\beta} D_{\alpha\beta} . \tag{2.1}$$

Here we have introduced the substantial time derivative:

$$\frac{\mathrm{D}}{\mathrm{D}t} \equiv \frac{\partial}{\partial t} + u_\alpha \frac{\partial}{\partial x_\alpha} , \tag{2.2}$$

and index notation with the usual summation convention has been used. The quantities appearing in these equations are [2]: $\rho(\boldsymbol{r},t)$ the mass density, $u_i(\boldsymbol{r},t)$ the local flow velocity, $\mathrm{e}(\boldsymbol{r},t)$ the local internal energy per unit mass, $P_{ij}(\boldsymbol{r},t)$ the pressure tensor, $q_i(\boldsymbol{r},t)$ the heat current density, and

$$D_{ij} \equiv \frac{1}{2}\left(\frac{\partial u_i}{\partial x_j} + \frac{\partial u_j}{\partial x_i}\right)$$

the rate of strain tensor. The five equations (2.1) are simply expressions of the conservation laws for number of particles, three components of linear momentum, and energy, and as such are rigorously and generally true. However, they become closed equations for the determination of ρ, $\boldsymbol{u}$ and e from their initial values only if the heat current density and the pressure tensor can be expressed in terms of these quantities, and the view of classical hydrodynamics is that these relations are given by phenomonological laws. Thus the heat current density is given by Fourier's law of heat conduction:

$$q_i = -\kappa \frac{\partial T}{\partial x_i}, \tag{2.3}$$

where κ is the heat conduction coefficient and T is the local temperature. The pressure tensor may be written

$$P_{ij} = p\delta_{ij} - \sigma_{ij}, \tag{2.4}$$

where p is the local hydrostatic pressure and σ_{ij} is the viscous stress tensor, given by a generalized Newton's law of friction:

$$\sigma_{ij} = 2\eta(D_{ij} - \tfrac{1}{3} D_{\alpha\alpha}\delta_{ij}) + \zeta D_{\alpha\alpha}\delta_{ij}. \tag{2.5}$$

Here η and ζ are the first and second coefficients of viscosity. In general it is found [3] that the transport coefficients κ, η, ζ are functions of T and ρ. Finally, it is assumed that the local quantities ρ, e, T and p are all related by the equilibrium thermodynamic equations of state for the fluid. Thus, for example, p and e are given as functions of ρ and T by the usual kinetic and caloric equations of state.

Since it is assumed that the usual thermodynamic relations apply between the various dependent variables, we can introduce the entropy per unit mass, s, which satisfies the well known thermodynamic relation:

$$T\,\mathrm{d}s = \mathrm{de} - \frac{p}{\rho^2}\,\mathrm{d}\rho \tag{2.6}$$

or

$$T\frac{\mathrm{D}s}{\mathrm{D}t} = \frac{\mathrm{De}}{\mathrm{D}t} - \frac{p}{\rho^2}\frac{\mathrm{D}\rho}{\mathrm{D}t}.$$

Using the hydrodynamic equations (2.1) we find

$$\rho T\frac{\mathrm{D}s}{\mathrm{D}t} = -\frac{\partial q_\alpha}{\partial x_\alpha} + \sigma_{\alpha\beta}D_{\alpha\beta}, \tag{2.7}$$

where (2.4) has been used. This equation replaces the last hydrodynamic equation of (2.1) and may be more familiar to some readers. It states that in the absence of viscosity and heat conduction, i.e. in an ideal fluid, the entropy per unit mass is constant along a streamline.

The fluid motion we call sound corresponds to a small disturbance from the quiescent or equilibrium state of the fluid, where all the thermodynamic quantities are constant in space and time. Hence it is appropriate to linearize the hydrodynamic equations by replacing each of these quantities by its equilibrium value plus a small fluctuation, and ignoring terms which are of second order in the fluctuations. Although it may not appear natural at this stage of the discussion, it will turn out to be convenient to choose the pressure and the entropy per unit mass as the thermodynamic variables, so we replace

$$p(\boldsymbol{r},t) \to p + p'(\boldsymbol{r},t) \ ,$$

$$s(\boldsymbol{r},t) \to s + s'(\boldsymbol{r},t) \ , \tag{2.8}$$

where the unprimed quantities are the equilibrium values. For simplicity we assume the equilibrium flow velocity is zero, so

$$\boldsymbol{u}(\boldsymbol{r},t) \to \boldsymbol{u}'(\boldsymbol{r},t) \ . \tag{2.9}$$

The density and temperature are expressed in terms of the pressure and the entropy through the equations of the state, so we replace

$$\rho(\boldsymbol{r},t) \to \rho + \left(\frac{\partial \rho}{\partial p}\right)_s p'(\boldsymbol{r},t) + \left(\frac{\partial \rho}{\partial s}\right)_p s'(\boldsymbol{r},t) \ ,$$

$$T(\boldsymbol{r},t) \to T + \left(\frac{\partial T}{\partial p}\right)_s p'(\boldsymbol{r},t) + \left(\frac{\partial T}{\partial s}\right)_p s'(\boldsymbol{r},t) \ . \tag{2.10}$$

Making these replacements in the hydrodynamic equations and keeping only those terms which are of first order in the primed quantities we obtain the *linearized hydrodynamic equations*:

$$\left(\frac{\partial \rho}{\partial s}\right)_p \frac{\partial s'}{\partial t} + \left(\frac{\partial \rho}{\partial p}\right)_s \frac{\partial p'}{\partial t} + \rho \operatorname{div} \boldsymbol{u}' = 0 \ ,$$

$$\rho \frac{\partial \boldsymbol{u}'}{\partial t} + \boldsymbol{\nabla} p' = \eta \nabla^2 \boldsymbol{u}' + (\zeta + \tfrac{1}{3}\eta) \boldsymbol{\nabla}(\boldsymbol{\nabla} \cdot \boldsymbol{u}') \ ,$$

$$\frac{\partial s'}{\partial t} = \frac{\kappa}{\rho T}\left(\frac{\partial T}{\partial s}\right)_p \nabla^2 s' + \frac{\kappa}{\rho T}\left(\frac{\partial T}{\partial p}\right)_s \nabla^2 p' \ . \tag{2.11}$$

The motion we call sound is a solution of these equations, as we can see most simply by first neglecting viscosity and heat conduction (ideal fluid approximation). If we then eliminate $\boldsymbol{u}'$ between the first two equations and use the third to eliminate s', we find that the pressure fluctuations satisfy the wave equation:

$$\frac{\partial^2 p'}{\partial t^2} - v_0^2 \nabla^2 p' = 0 , \tag{2.12}$$

where

$$v_0 = \sqrt{\left(\frac{\partial p}{\partial \rho}\right)_s} \tag{2.13}$$

is the well known Laplace expression for the velocity of sound in an ideal fluid in terms of the adiabatic compressibility.

2. THE NORMAL MODE SOLUTIONS

In order to investigate the character of the solutions of the linearized hydrodynamic equations when the effects of viscosity and heat conduction are included we consider the general plane wave or *normal mode* solution. That is, we write

$$p'(\boldsymbol{r},t) = \pi\, \mathrm{e}^{i(\boldsymbol{k}\cdot\boldsymbol{r} - \omega t)} ,$$

$$s'(\boldsymbol{r},t) = \sigma\, \mathrm{e}^{i(\boldsymbol{k}\cdot\boldsymbol{r} - \omega t)} ,$$

$$\boldsymbol{u}'(\boldsymbol{r},t) = \boldsymbol{\varphi}\, \mathrm{e}^{i(\boldsymbol{k}\cdot\boldsymbol{r} - \omega t)} , \tag{2.14}$$

and determine the five complex amplitudes π, σ, $\boldsymbol{\varphi}$ by inserting these expressions in the equations (2.11). The result is a set of five coupled homogeneous linear equations:

$$\omega\left(\frac{\partial \rho}{\partial s}\right)_p \sigma + \omega\left(\frac{\partial \rho}{\partial p}\right)_s \pi - \rho\boldsymbol{k}\cdot\boldsymbol{\varphi} = 0 , \tag{2.15a}$$

$$\rho\omega\boldsymbol{\varphi} - \boldsymbol{k}\pi = -ik^2\eta\,\boldsymbol{\varphi} - i(\tfrac{1}{3}\eta + \zeta)\boldsymbol{k}(\boldsymbol{k}\cdot\boldsymbol{\varphi}) , \tag{2.15b}$$

$$\omega\sigma = -ik^2 \frac{\kappa}{\rho T}\left(\frac{\partial T}{\partial s}\right)_p \sigma - ik^2 \frac{\kappa}{\rho T}\left(\frac{\partial T}{\partial p}\right)_s \pi . \tag{2.15c}$$

These equations simplify somewhat if we write

$$\boldsymbol{\varphi} = \boldsymbol{\varphi}^{\perp} + \varphi^{\parallel}\,\hat{\boldsymbol{k}}, \qquad \boldsymbol{k}\cdot\boldsymbol{\varphi}^{\perp} = 0 , \tag{2.16}$$

which corresponds to the separation of the velocity field into longitudinal and transverse parts [4]. By inspection of (2.15b) we see that the equation for the transverse velocity amplitude is

$$(\rho\omega + ik^2\eta)\boldsymbol{\varphi}^{\perp} = 0 \ , \tag{2.17}$$

while the remaining amplitudes fulfill the equations:

$$\omega\left(\frac{\partial\rho}{\partial s}\right)_p \sigma + \omega\left(\frac{\partial\rho}{\partial p}\right)_s \pi - \rho k \,\varphi^{\parallel} = 0 \ ,$$

$$-k\pi + [\rho\omega + ik^2(\tfrac{4}{3}\eta + \zeta)]\,\varphi^{\parallel} = 0 \ ,$$

$$\left[\omega + ik^2 \frac{\kappa}{\rho T}\left(\frac{\partial T}{\partial s}\right)_p\right]\sigma + ik^2\frac{\kappa}{\rho T}\left(\frac{\partial T}{\partial p}\right)_s \pi = 0 \ . \tag{2.18}$$

Equation (2.17) is really two equations for the two components of $\boldsymbol{\varphi}$ perpendicular to $\boldsymbol{k}$, i.e. the two polarizations of the transverse wave. A non-trivial solution exists if and only if

$$\omega = -i\frac{\eta}{\rho}k^2 \ , \tag{2.19}$$

which is the dispersion law for the two normal modes in which the transverse velocity oscillates. These are *shear waves* and the fact that for real k the frequency is pure imaginary tells us that such waves do not propagate; an initial shear wave disturbance relaxes monotonically in time toward equilibrium.

A non-trivial solution of the three coupled equations (2.18) exists if and only if the determinant of the coefficients vanishes. Setting this determinant equal to zero gives us

$$\left[\omega^2 - k^2\left(\frac{\partial p}{\partial\rho}\right)_s + i\omega\frac{k^2}{\rho}(\tfrac{4}{3}\eta+\zeta)\right]\left[\omega + ik^2\frac{\kappa}{\rho T}\left(\frac{\partial T}{\partial s}\right)_p\right]$$
$$-i\omega k^2\frac{\kappa}{\rho T}\left(\frac{\partial T}{\partial p}\right)_s\left(\frac{\partial\rho}{\partial s}\right)_p\left(\frac{\partial p}{\partial\rho}\right)_s\left[\omega + i\frac{k^2}{\rho}(\tfrac{4}{3}\eta+\zeta)\right] = 0 \ . \tag{2.20}$$

This equation can be put in a more perspicuous form if we use some simple thermodynamic relations. Thus,

$$T\left(\frac{\partial s}{\partial T}\right)_p = c_p \tag{2.21}$$

is the specific heat at constant pressure. Since

$$\left(\frac{\partial T}{\partial s}\right)_p = \left(\frac{\partial T}{\partial s}\right)_\rho + \left(\frac{\partial T}{\partial\rho}\right)_s\left(\frac{\partial\rho}{\partial s}\right)_p, \quad \left(\frac{\partial T}{\partial\rho}\right)_s = \left(\frac{\partial T}{\partial p}\right)_s\left(\frac{\partial p}{\partial\rho}\right)_s \ ,$$

we have

$$\left(\frac{\partial T}{\partial p}\right)_s \left(\frac{\partial \rho}{\partial s}\right)_p \left(\frac{\partial p}{\partial \rho}\right)_s = \left(\frac{\partial T}{\partial s}\right)_p - \left(\frac{\partial T}{\partial s}\right)_\rho = -T\left(\frac{1}{c_v} - \frac{1}{c_p}\right) , \quad (2.22)$$

where c_v is the specific heat at constant volume (or density). Using these relations we can write (2.20) in the form:

$$\left(\omega + ik^2 \frac{\kappa}{\rho c_p}\right)\left\{\omega^2 - k^2 v_0^2 + i\omega \frac{k^2}{\rho}\left[\tfrac{4}{3}\eta + \zeta + \kappa\left(\frac{1}{c_v} - \frac{1}{c_p}\right)\right]\right\}$$

$$- \omega k^4 \frac{\kappa}{\rho^2}\left(\frac{1}{c_v} - \frac{1}{c_p}\right)\left(\tfrac{4}{3}\eta + \zeta - \frac{\kappa}{c_p}\right) = 0 , \quad (2.23)$$

where v_0 is the ideal fluid sound velocity, given by (2.13). This equation relating ω and k is the dispersion law for the three normal modes in which the longitudinal velocity, pressure, and entropy fluctuate. If we consider k to be given then it is a cubic equation for ω as a function of k, so there are three roots with three corresponding normal mode solutions. The explicit form of these roots can, of course, be obtained. We are at the moment, however, interested primarily in the character of the solutions of the hydrodynamic equations when the effects of viscosity and heat conduction are small, and clearly this will be the case whenever k, the magnitude of the propagation vector, is small [5]. Since $k = 2\pi/\lambda$, where λ is the wave-length of the disturbance, small k corresponds to long wave-length, i.e. disturbances whose spatial variation is small.

For small k, the solutions of (2.23) can be expanded in powers of k. We find one imaginary root:

$$\omega = -i\left[\frac{\kappa}{\rho c_p}k^2 - \frac{\kappa^2}{\rho^3 v_0^2 c_p}\left(\frac{1}{c_v} - \frac{1}{c_p}\right)\left(\tfrac{4}{3}\eta + \zeta - \frac{\kappa}{c_p}\right)k^4 + \dots\right], \quad (2.24)$$

and two roots with equal imaginary parts and with real parts of opposite sign:

$$\omega = \pm v_0 k - i\frac{1}{2\rho}\left[\tfrac{4}{3}\eta + \zeta + \kappa\left(\frac{1}{c_v} - \frac{1}{c_p}\right)\right]k^2$$

$$\mp \frac{1}{8\rho^2 v_0}\left\{\left[\tfrac{4}{3}\eta + \zeta + \kappa\left(\frac{1}{c_v} - \frac{1}{c_p}\right)\right]^2 - 4\kappa\left(\frac{1}{c_v} - \frac{1}{c_p}\right)\left(\tfrac{4}{3}\eta + \zeta - \frac{\kappa}{c_p}\right)\right\}k^3$$

$$- i\frac{\kappa^2}{2\rho^3 v_0^2 c_p}\left(\frac{1}{c_v} - \frac{1}{c_p}\right)\left(\tfrac{4}{3}\eta + \zeta - \frac{\kappa}{c_p}\right)k^4 + \dots \quad (2.25)$$

This last pair of roots corresponds to the sound mode, the duality arising from the two senses of propagation. The leading term of the imaginary part, which gives the decay in time of long wavelength sound, is the famous result of Kirchhoff [6)], obtained in 1868.

The normal mode solutions of the coupled equations (2.18) are obtained by using the dispersion laws (2.24) and (2.25) for the respective modes and then solving for the fluctuation amplitudes. Thus, inserting the expression (2.24) for ω in (2.18) we find the soluton to be

$$\sigma = \text{arbitrary complex amplitude,}$$

$$\pi \approx 0,$$

$$\varphi'' \approx -ik\left(\frac{\partial\rho}{\partial s}\right)_p \frac{\kappa}{\rho^2 c_p}\,\sigma\,, \tag{2.26}$$

where we have neglected terms of order k^2. Thus we see that, for small k, this mode is primarily an entropy, or heat conduction, mode. The fluctuations of the longitudinal velocity are small compared with the entropy fluctuations while the pressure fluctuations are still smaller.

Consider next the sound modes, for which the dispersion law is (2.25). Inserting this relation in (2.18) we find the solution to be:

$$\pi = \text{arbitrary complex amplitude,}$$

$$\varphi'' \approx \left\{\pm\frac{1}{\rho v_0} - i\,\frac{k}{2\rho^2 v_0^2}\left[\tfrac{4}{3}\eta + \zeta - \kappa\left(\frac{1}{c_v} - \frac{1}{c_p}\right)\right] + \dots\right\}\pi,$$

$$\sigma \approx \pm ik\,\frac{\kappa}{\rho v_0 T}\left(\frac{\partial T}{\partial p}\right)_s \pi\,, \tag{2.27}$$

where, again, terms of order k^2 have been neglected. For these modes, we see, the fluctuations in pressure and longitudinal velocity are of the same order while the entropy fluctuations are small. This is in complete accord with our elementary notions of sound as a propagating pressure-velocity disturbance.

We conclude this section with a number of remarks:

i) We have found exactly five normal mode solutions of the linearized hydrodynamic equations. This is clearly a consequence of the fact that there are five hydrodynamic equations, each first order in time, and that we have asked for the normal modes for fixed propagation vector $\boldsymbol{k}$. For each mode the dispersion law is an expression for ω as a function of k.

ii) For small k the various modes each have a distinctive character. Thus the shear modes involve only fluctuations of transverse velocity, the heat conduction mode involves primarily fluctuations in entropy, while the sound mode is primarily a pressure and longitudinal velocity mode. This has the consequence, for example, that at low frequencies a vibrating tuning fork will excite only the sound mode since it creates primarily a pressure disturbance. At high frequencies (or short wave lengths) the sharp distinction between the character of the sound and heat conduction modes does not persist; they each involve fluctuations of pressure, entropy, and longitudinal velocity which are all of the same order of magnitude.

iii) This last remark takes on added force when we consider ω real and solve for complex k, which, as we remarked in the introduction, is appropriate for the source problem. There are now a *pair* of heat mode roots:

$$k = \pm (1+i)\left(\frac{\rho c_p \omega}{2\kappa}\right)^{\frac{1}{2}} \left[1 + i \frac{c_p}{2\rho v_0^2 \kappa}\left(\frac{1}{c_v} - \frac{1}{c_p}\right)\left(\tfrac{4}{3}\eta + \zeta - \frac{\kappa}{c_p}\right)\omega + \dots\right] ,$$

again with the duality corresponding to the direction in which the wave travels. The important thing to note here is that this mode has a real phase velocity:

$$v_{\mathrm{ph}} = \left|\mathrm{Re}\left\{\frac{k}{\omega}\right\}\right|^{-1}$$

$$= \left(\frac{2\kappa\omega}{\rho c_p}\right)^{\frac{1}{2}} \left[1 + \frac{c_p}{2\rho v_0^2 \kappa}\left(\frac{1}{c_v} - \frac{1}{c_p}\right)\left(\tfrac{4}{3}\eta + \zeta - \frac{\kappa}{c_p}\right)\omega + \dots\right],$$

and, hence, in this respect at least, has the characteristics of a propagating mode. We conclude that the heat conduction mode as well as the sound mode can be excited by a very high frequency source and that the propagation characteristics of the two modes will be similar.

iv) We can in part understand from the macroscopic point of view why the absorption of sound increases with increasing frequency. The point is that the characteristic time for dissipative effects to occur is proportional to the square of the spatial gradients (a typical time is $\rho L^2/\eta$, where L is a characteristic length). Thus, *although* as the frequency increases there is less time for dissipative effects to influence propagation, the effects of spatial gradients dominate as the wave length decreases and absorption increases.

3. CASE OF THE IDEAL MONOATOMIC GAS, THE SIMPLEST RESULTS OF KINETIC THEORY

What happens as the wave length of sound is made shorter and shorter so the effects of viscosity and heat conduction are no longer small? The validity of the hydrodynamic equations themselves becomes questionable under such conditions, as we shall see more explicitly in the next section, but suppose for the moment we consider them as given. What we must do then is discuss the dispersion law (2.23) for arbitrary k. This discussion is made much simpler if we specialize to the case of the ideal monoatomic gas, which is in any event the case of real interest to us in this work.

For an ideal gas the equation of state is

$$p = \rho \mathcal{k} T/m \ , \tag{2.28}$$

where m is the molecular mass, and $\mathcal{k}$ is Boltzmann's constant, not to be confused with the magnitude of the propagation vector. For a monoatomic ideal gas the specific heats are

$$c_v = 3\mathcal{k}/2m, \qquad c_p = 5\mathcal{k}/2m \ . \tag{2.29}$$

Since in general

$$\left(\frac{\partial p}{\partial \rho}\right)_S = \frac{c_p}{c_v}\left(\frac{\partial p}{\partial \rho}\right)_T \ , \tag{2.30}$$

we have from (2.13) the well known expression for the velocity of sound in an ideal monoatomic gas:

$$v_0 = \sqrt{\frac{5}{3}\frac{\mathcal{k}T}{m}} \ . \tag{2.31}$$

For an ideal monoatomic gas the second viscosity coefficient ζ must vanish. This follows from the fact that the *hydrodynamic pressure:*

$$P_{\text{hyd}} \equiv \tfrac{1}{3} P_{\alpha\alpha} = p - \zeta \operatorname{div} \boldsymbol{u} \ , \tag{2.32}$$

differs from the hydrostatic pressure p by a term proportional to ζ. But from kinetic theory we know that both pressures are given by the ideal gas equation (2.28), the difference being that for the hydrodynamic pressure the temperature is that associated with translational kinetic energy alone, while for the hydrostatic pressure the temperature is that associated with both translational and internal degrees of freedom of the gas molecules. Since there are no internal degrees of freedom for monoatomic gases these two temperatures and, hence, the corresponding pressures must be equal, and ζ must vanish.

The dimensionless ratio

$$f = \frac{\kappa}{c_v \eta} = \frac{2\, m\kappa}{3\, k\eta} \tag{2.33}$$

is called the Eucken number and is found both experimentally and theoretically to be nearly a constant with the same value $\frac{5}{2}$ for all monoatomic gases [7].

Finally the quantity

$$\Lambda \equiv \frac{\eta}{\rho v_0}, \tag{2.34}$$

which has the dimensions of a length, may be identified as the mean free path of the gas molecules [8]. The dispersion law (2.23) may be now written in the form:

$$\left(\xi + \tfrac{3}{5}\, i f x^2\right)\left[\xi^2 - x^2 + i\left(\tfrac{4}{3} + \tfrac{2}{5}\, f\right)\xi x^2\right] + \tfrac{2}{5} f \left(\tfrac{3}{5} f - \tfrac{4}{3}\right)\xi x^4 = 0 \,, \tag{2.35}$$

where

$$\xi = \frac{\omega\Lambda}{v_0}, \qquad x = k\Lambda \,. \tag{2.36}$$

The parameter x is the ratio of the mean free path to the wave length, while ξ is the ratio of the frequency to the collision frequency v_0/Λ. We call ξ the dimensionless frequency and x the dimensionless wave number. When the wave length is long compared with the mean free path, x is small and the roots of (2.35) are given by (2.24) and (2.25) specialized to the ideal gas case. For still larger values of x it is not much trouble to determine ξ numercially from (2.35), and we have done so. The root corresponding to the pure imaginary "heat conduction" mode is shown in fig. 1. The magnitude of this root increases monotonically with increasing x, indicating that the corresponding mode is more and more highly damped as the wave length gets shorter and shorter. The behaviour of the sound mode roots is more interesting, as shown in fig. 2. The real part of ξ/x, which is the ratio of the phase velocity of the wave to the Laplace velocity, diminishes monotonically until it vanishes at $x = x_0 = 0.979...$

If we remember that the sound mode roots are a pair of roots

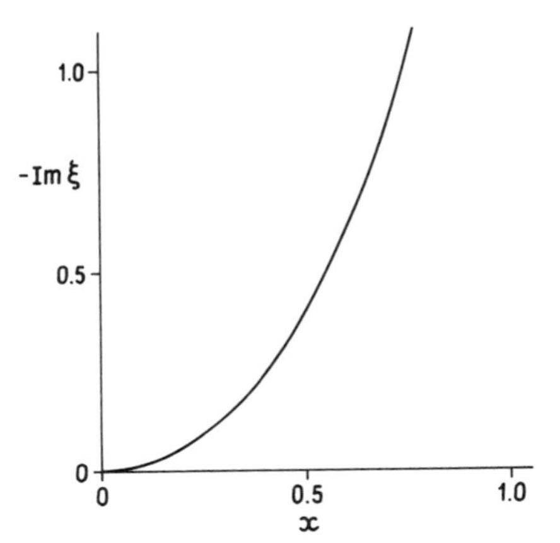

Fig. 1. The imaginary part of the dimensionless frequency as a function of the dimensionless wave number for the heat conduction mode from the Navier-Stokes equations.

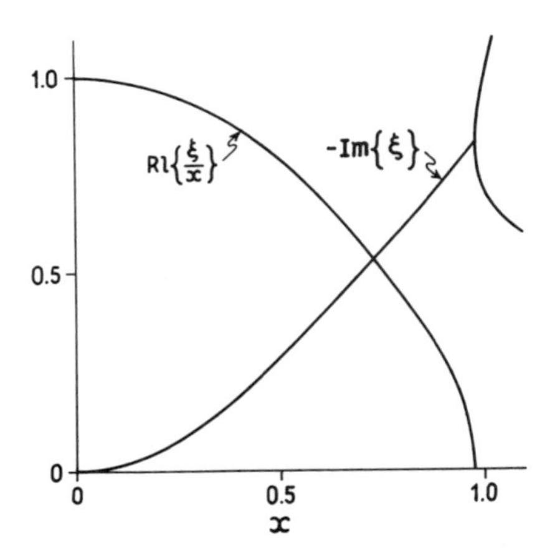

Fig. 2. The real and imaginary parts of the dimensionless frequency as functions of the dimensionless wave number for the sound modes from the Navier-Stokes equations.

with equal imaginary parts and with real parts of opposite sign, we see that at this value of x there is a double root of the dispersion law. For still larger x these two "sound mode" roots remain pure imaginary, one increasing in magnitude for increasing x, the other decreasing toward an asymptotic value.

The character of the modes themselves is determined from the set of equations (2.18). Using the properties of the ideal gas we have introduced, together with some simple thermodynamics [9], we can put these equations in the form:

$$-\xi \frac{\sigma}{c_p} + \xi \frac{\pi}{\rho v_0^2} - x \frac{\varphi^{\shortparallel}}{v_0} = 0 \ ,$$

$$-x \frac{\pi}{\rho v_0^2} + \left(\xi + \tfrac{4}{3} i\, x^2\right) \frac{\varphi^{\shortparallel}}{v_0} = 0 \ ,$$

$$\left(\xi + \tfrac{3}{5} if\, x^2\right) \frac{\sigma}{c_p} + \tfrac{2}{5} if\, x^2 \frac{\pi}{\rho v_0^2} = 0 \ . \tag{2.37}$$

Thus, for example, when $x = 0.6$, $\xi = 0.41 - 0.40\, i$ for the sound mode, and the solution of these equations is

$$\pi = \text{arbitrary complex amplitude,}$$

$$\sigma \approx -(0.33 + 0.74\, i) \frac{c_p}{\rho v_0^2} \pi \ ,$$

$$\varphi \approx (1.4 - 0.25\, i) \frac{1}{\rho v_0} \pi \ .$$

On the other hand for the same value of x, $\xi = -0.6\, i$ for the heat conduction mode, and the solution of these equations is

$$\pi = \text{arbitrary complex amplitude,}$$

$$\sigma = +6i \frac{c_p}{\rho v_0^2} \pi \ ,$$

$$\varphi = -5i \frac{\pi}{\rho v_0} \ .$$

Here we see explicitly that when the mean free path is not small compared with the wave length, the character of the motion in the sound and heat conduction modes becomes similar. In particular, the pressure fluctuates in both these modes.

Of course, as we remarked above, at low frequencies the wave length is long compared with the mean free path and we can expand the roots ξ of the dispersion law (2.35) in powers of x. The result for the sound mode is

$$\xi = x - i\left(\tfrac{2}{3} + \tfrac{1}{5}f\right)x^2 - \left(\tfrac{2}{9} - \tfrac{2}{15}f + \tfrac{7}{50}f^2\right)x^3 + i\,\tfrac{3}{25}f^2\left(\tfrac{3}{5}f - \tfrac{4}{3}\right)x^4 + \dots\,, \tag{2.38}$$

where we have chosen the sound mode root whose expansion begins with $+x$; the other sound mode root is obtained by replacing x by $-x$ in this expression.

For comparison with experiments in which the phase velocity and absorption length are measured as functions of the frequency, it is necessary to invert the series (2.38) to obtain the expansion:

$$x = \xi + i\left(\tfrac{2}{3} + \tfrac{1}{5}f\right)\xi^2 - \left(\tfrac{2}{3} + \tfrac{2}{3}f - \tfrac{3}{50}f^2\right)\xi^3$$

$$- i\left(\tfrac{20}{27} + \tfrac{14}{9}f - \tfrac{7}{75}f^2 - \tfrac{7}{250}f^3\right)\xi^4 + \dots\,. \tag{2.39}$$

Then, for real ω, the phase velocity and absorption length are given by (1.3) and (1.4). Referring to (2.36) we write:

$$\frac{v_o}{v_{ph}} = \mathrm{Re}\,\frac{x}{\xi} = 1 - \left(\tfrac{2}{3} + \tfrac{2}{3}f - \tfrac{3}{50}f^2\right)\xi^2 + \dots\,, \tag{2.40a}$$

$$\frac{\alpha v_o}{\omega} = \mathrm{Im}\,\frac{x}{\xi} = \left(\tfrac{2}{3} + \tfrac{1}{5}f\right)\xi - \left(\tfrac{20}{27} + \tfrac{14}{9}f - \tfrac{7}{75}f^2 - \tfrac{7}{250}f^3\right)\xi^3 + \dots\,. \tag{2.40b}$$

We shall return to these expressions in chapter IV.

4. THE CHAPMAN-ENSKOG DEVELOPMENT, THE BURNETT EQUATIONS

Up to now we have adopted the view that the equations of hydrodynamics are phenomonological equations and have investigated their consequences without questioning their validity. It is well known, however, that these equations can be derived from kinetic theory. What we have in mind is the Chapman-Enskog successive approximation scheme for the solution of the Boltzmann transport equation [10]. In this section we shall briefly describe the nature of the Chapman-Enskog development and the resulting hydrodynamic equations. Then we describe their consequences when applied to the sound dispersion problem.

The Chapman-Enskog development is an approximation scheme for the determination of the molecular distribution function $f(\boldsymbol{r},\boldsymbol{v},t)$, where $f(\boldsymbol{r},\boldsymbol{v},t)\,\mathrm{d}\boldsymbol{r}\,\mathrm{d}\boldsymbol{v}$ is the number of molecules of the gas whose position [11] is in a cell $\mathrm{d}\boldsymbol{r} = \mathrm{d}x\,\mathrm{d}y\,\mathrm{d}z$ about $\boldsymbol{r}$ and whose velocity is in a cell $\mathrm{d}\boldsymbol{v} = \mathrm{d}v_x\,\mathrm{d}v_y\,\mathrm{d}v_z$ about $\boldsymbol{v}$. The various quantities introduced in section 1 are expressible as simple averages of this distribution function:

$$\rho(\boldsymbol{r},t) = m\int \mathrm{d}\boldsymbol{v}\, f(\boldsymbol{r},\boldsymbol{v},t) \;,$$

$$\rho u_i(\boldsymbol{r},t) = m\int \mathrm{d}\boldsymbol{v}\, v_i\, f(\boldsymbol{r},\boldsymbol{v},t) \;,$$

$$\rho \mathrm{e}(\boldsymbol{r},t) = \tfrac{1}{2} m\int \mathrm{d}\boldsymbol{v}\,(\boldsymbol{v}-\boldsymbol{u})^2 f(\boldsymbol{r},\boldsymbol{v},t) \;,$$

$$P_{ij}(\boldsymbol{r},t) = m\int \mathrm{d}\boldsymbol{v}\,(v_i-u_i)(v_j-u_j)\, f(\boldsymbol{r},\boldsymbol{v},t) \;,$$

$$q_i(\boldsymbol{r},t) = \tfrac{1}{2} m\int \mathrm{d}\boldsymbol{v}\,(v_i-u_i)(\boldsymbol{v}-\boldsymbol{u})^2 f(\boldsymbol{r},\boldsymbol{v},t) \;. \qquad (2.41)$$

The starting point of the Chapman-Enskog scheme is to take $f(\boldsymbol{r},\boldsymbol{v},t)$ to be a local Maxwellian distribution:

$$f^{(0)}(\boldsymbol{r},\boldsymbol{v},t) = \frac{\rho(\boldsymbol{r},t)}{m}\left[\frac{m}{2\pi k T(\boldsymbol{r},t)}\right]^{\frac{3}{2}} \exp\left\{-\frac{m[\boldsymbol{v}-\boldsymbol{u}(\boldsymbol{r},t)]^2}{2 k T(\boldsymbol{r},t)}\right\} \;. \qquad (2.42)$$

If this is inserted in (2.41) one finds $\mathrm{e} = \frac{3}{2} kT/m$, $q_i^{(0)} = 0$, and $P_{ij}^{(0)} = p\delta_{ij}$ with $p = \rho kT/m$. That is, one obtains the ideal fluid equations of hydrodynamics with the ideal gas equation for the thermodynamic equation of state. Of course, $f^{(0)}$ is a solution of the Boltzmann equation only if ρ, T, and $\boldsymbol{u}$ are constants, so in the next approximation where one writes

$$f(\boldsymbol{r},\boldsymbol{v},t) = f^{(0)}[1 + h(\boldsymbol{r},\boldsymbol{v},t)] \;, \qquad (2.43)$$

and takes h to be small, it is clear that h will be proportional to the derivatives of ρ, $\boldsymbol{u}$ and T with respect to time and space. Actually, it turns out that the derivatives with respect to time are eliminated by using the ideal fluid equations and h depends only upon the spatial gradiants. When this result is inserted in (2.41) one obtains the Fourier law of heat conduction and the Newton law of friction, i.e. equations (2.3) and (2.5), but with explicit expressions for the kinetic coefficients. The second viscosity coefficient vanishes, as it must for an ideal monatomic gas, while the expressions for κ and η depend upon the molecular force law. Thus, for elastic spheres of diameter a, one finds [12]:

$$\eta = 1.016 \frac{5m}{16a^2}\left(\frac{kT}{\pi m}\right)^{\frac{1}{2}}, \qquad \kappa = 1.025 \frac{75k}{64a^2}\left(\frac{kT}{\pi m}\right)^{\frac{1}{2}} . \tag{2.44}$$

For Maxwell molecules, for which the intermolecular potentials is of the form: $\varphi(r) = \varphi_4(a/r)^4$, where r is the intermolecular distance, one finds [13]:

$$\eta = 2.29 \left(\frac{m}{2\varphi_4}\right)^{\frac{1}{2}} \frac{kT}{3\pi a^2}, \qquad \kappa = 2.29 \left(\frac{m}{2\varphi_4}\right)^{\frac{1}{2}} \frac{5k^2 T}{4\pi m a^2} . \tag{2.45}$$

Note that the Eucken constant, given by (2.33), has exactly the value 2.5 for Maxwell molecules, while it has the value 2.522... for hard spheres. Other molecular models give values lying between these two values [14]. Note, also that for hard spheres the mean free path defined in (2.34) is

$$\Lambda = 1.016 \left(\frac{3}{5\pi}\right)^{\frac{1}{2}} \frac{5m}{16\rho a^2}, \tag{2.46}$$

which is, aside from a numerical factor, exactly the classical definition of the mean free path [8].

What happens when one goes to higher approximations? The result is a sequence of corrections to the Fourier and Newton laws. One finds

$$\begin{aligned} P_{ij} &= p\,\delta_{ij} + P_{ij}^{(1)} + P_{ij}^{(2)} + \dots , \\ q_i &= q_i^{(1)} + q_i^{(2)} + \dots , \end{aligned} \tag{2.47}$$

where $q_i^{(1)}$ is given by (2.3) and $P_{ij}^{(1)}$ (with a sign change) is given by (2.5). The second approximations $q^{(2)}$ and $P_{ij}^{(2)}$ are given in the book of Chapman and Cowling; the resulting hydrodynamics equations are termed the Burnett equations [15]. The successive approximations, as one might expect, involve successively higher spatial derivatives, and the expansion is essentially an expansion in $\Lambda\,\partial/\partial \boldsymbol{r}$, i.e. the relative change of the macroscopic variables over a mean free path. Thus, in the applications to the sound propagation problem, the expansion parameter is the ratio of the mean free path to the wave length.

Here we see the heart of the problem. Any attempt to discuss the dispersion of sound beyond the classical result of Kirchhoff involves the discussion of terms of higher power than the first in the ratio of the mean free path to the wave length, but it is exactly such terms which arise in corrections to the hydrodynamic equations. Hence, one cannot avoid the problem of the solution of the Boltzmann equation, either by means of the Chapman-Enskog scheme, or by other means.

The discussion of sound dispersion for the Burnett equations was first given by Primakoff, and later by Tsien and Schamberg [16]. Unfortunately, the expression for $q_i^{(2)}$, which these authors took from the first edition of the book of Chapman and Cowling, was incorrect, which means that their expressions for the sound dispersion are in error. The error was found and corrected by C.S. Wang Chang and G. E. Uhlenbeck, who discussed also the dispersion law resulting from the third approximation to the hydrodynamic equations [17]. The dispersion law obtained by them from the Burnett equations is

$$(\xi + i\,\tfrac{3}{5}f\;x^2)\,[\xi^2 - x^2 + i(\tfrac{4}{3} + \tfrac{2}{5}f)\xi x^2] + [\tfrac{2}{5}f(\tfrac{3}{5}f - \tfrac{4}{3}) + \tfrac{16}{225}f^2 C_1 + \tfrac{16}{45}f\,C_2 - \tfrac{10}{9}C_3]\xi x^4 = 0\;, \tag{2.48}$$

where f is the Eucken number defined in (2.33) and the parameters C_1, C_2 and C_3 are dimensionless quantities which depend upon the force law. For molecules which repel each other with a force proportional to an inverse power of the distance these parameters are pure numbers; otherwise they are weakly dependent upon the temperature. In table 1 the value of these parameters for several molecular models is reproduced from the paper of Chang [18].

Table 1
The parameters appearing in the dispersion law for the Burnett equations

	Maxwell molecules	Hard spheres	Helium molecules
$-\frac{8}{45}C_1$	1	1.017	1.008
$\frac{1}{3}C_2$	1	0.800	0.855
$\frac{1}{2}C_3$	1	1.014	1.007
$\frac{2}{5}f$	1	1.009	1.004

Note that the Burnett dispersion law (2.48) is still a cubic equation in ξ, the dimensionless frequency, just as is the Navier-Stokes dispersion law. There will therefore again be three roots, one corresponding to the heat mode and two corresponding to the sound mode. If we expand in powers of x we find for the heat mode:

$$\xi = -i\,\tfrac{3}{5} f x^2 - i\,\tfrac{3}{5} f\,[\tfrac{2}{5} f(\tfrac{3}{5} f - \tfrac{4}{3}) + \tfrac{16}{225} f^2 C_1 + \tfrac{16}{45} f C_2 - \tfrac{10}{9} C_3]\,x^4 + \dots, \tag{2.49}$$

while for the sound modes:

$$\xi = \pm x - i(\tfrac{2}{3} + \tfrac{1}{5} f)x^2 \mp (\tfrac{2}{9} - \tfrac{2}{15} f + \tfrac{7}{50} f^2 + \tfrac{8}{225} f^2 C_1 + \tfrac{8}{45} f\, C_2 - \tfrac{5}{9} C_3)x^3$$
$$+ i\,\tfrac{3}{5} f\,[\tfrac{1}{5} f(\tfrac{3}{5} f - \tfrac{4}{3}) + \tfrac{8}{225} f^2 C_1 + \tfrac{8}{45} f C_2 - \tfrac{5}{9} C_3]\,x^4 + \dots . \tag{2.50}$$

The Navier-Stokes results (2.38) are recovered if we set $C_1 = C_2 = C_3 = 0$.

For comparison with experimental measurements of the dispersion and absorption of sound as functions of the frequency, it is necessary to invert the series (2.50) to express the dimensionless wave number x in powers of the dimensionless frequency ξ. Choosing the root with positive real part, we find:

$$x = \xi + i(\tfrac{2}{3} + \tfrac{1}{5} f)\xi^2 - (\tfrac{2}{3} + \tfrac{2}{3} f - \tfrac{3}{50} f^2 - \tfrac{8}{225} f^2 C_1 - \tfrac{8}{45} f C_2 + \tfrac{5}{9} C_3)\,\xi^3$$
$$- i\,[\tfrac{20}{27} + \tfrac{14}{9} f - \tfrac{7}{75} f^2 - \tfrac{7}{250} f^3 - (\tfrac{10}{3} + \tfrac{2}{5} f)(\tfrac{8}{225} f^2 C_1 + \tfrac{8}{45} f C_2 - \tfrac{5}{9} C_3]\,\xi^4 + \dots . \tag{2.51}$$

Referring to (1.3) and (1.4) we can write

$$\frac{v_0}{v_{ph}} = \mathrm{Re}\left\{\frac{x}{\xi}\right\} = 1 - (\tfrac{2}{3} + \tfrac{2}{3} f - \tfrac{3}{50} f^2 - \tfrac{8}{225} f^2 C_1 - \tfrac{8}{45} f C_2 + \tfrac{5}{9} C_3)\xi^2 + \dots \tag{2.52}$$

and

$$\frac{\alpha v_0}{\omega} = \mathrm{Im}\left\{\frac{x}{\xi}\right\} = (\tfrac{2}{3} + \tfrac{1}{5} f)\,\xi \tag{2.53}$$
$$- [\tfrac{20}{27} + \tfrac{14}{9} f - \tfrac{7}{75} f^2 - \tfrac{7}{250} f^3 - (\tfrac{10}{3} + \tfrac{2}{5} f)(\tfrac{8}{225} f^2 C_1 + \tfrac{8}{45} f\, C_2 - \tfrac{5}{9} C_3)]\,\xi^3 + \dots$$

To gain an idea of the size of the corrections to the Navier-Stokes dispersion, we note for the case of Maxwell molecules the expansions (2.52) and (2.53) become:

$$\frac{v_0}{v_{ph}} = 1 - \frac{215}{72}\,\xi^2 + \dots\ , \qquad (2.54a)$$

$$\frac{\alpha v_0}{\omega} = \frac{7}{6}\,\xi - \frac{3483}{432}\,\xi^3 + \dots\ , \qquad (2.54b)$$

while, setting $f = \frac{5}{2}$, the corresponding Navier-Stokes results (2.40) become:

$$\frac{v_0}{v_{ph}} = 1 - \frac{141}{72}\,\xi^2 + \dots\ , \qquad (2.55a)$$

$$\frac{\alpha v_0}{\omega} = \frac{7}{6}\,\xi - \frac{1559}{432}\,\xi^3 + \dots\ . \qquad (2.55b)$$

Clearly the corrections are significant!

We close with some remarks about the higher approximations in the Chapman-Enskog scheme.

a) In each order the dispersion law is a cubic equation in the frequency. Thus there are always three roots, corresponding to one heat conduction mode and two sound modes. In addition there are always two transverse sound modes. The existence of five normal modes is clearly a consequence of the fact that the five hydrodynamic equations (2.1) are first order in time, although in the Chapman-Enskog scheme the pressure tensor and the heat current density involve indefinitely high powers of the spatial gradients. As we have remarked, the form of the hydrodynamic equations is a consequence of the existence of five conservation laws: number of particles, energy, and three components of linear momentum. Thus, there is a deep connection between the hydrodynamic normal modes and the conservation laws.

b) In the Burnett approximation the first two terms in the expansion (2.50) of the dispersion law, the terms corresponding to Laplace dispersion and Kirchhoff absorption, are identical with the corresponding terms in the expansion (2.38) obtained in the Navier-Stokes approximation. In the next approximation, the so called super-Burnett approximation, the first three terms in the expansion of the dispersion law coincide with the first three terms obtained in the Burnett approximation. And so on. Thus in the n'th approximation the first $n + 1$ terms of the expansion of the dispersion law are determined; higher approximations only modify the higher order coefficients.

c) That there are only five normal modes in the hydrodynamic description of a gas has been called the Hilbert paradox. That is, we know that a gas consists of an indefinitely large number of particles and therefore has an indefinitely large number of degrees of freedom. But mechanical systems in general have as many normal modes of oscillation as they have degrees of freedom; how can it be that for a gas there are only five? We shall have something to say about this in chapter V.

NOTES

1) For a derivation of these equations from a kinetic point of view see G. E. Uhlenbeck and G. W. Ford, *Lectures in statistical Mechanics* (American Math. Soc., Providence, R. I., 1963) chapter VI.

2) The pressure tensor and the heat flux vector are defined as follows. Consider an infinitesimal element of surface in the fluid. Let $\mathrm{d}\boldsymbol{S}$ be an infinitesimal vector whose magnitude is equal to the area of the element and whose direction is perpendicular to the element of surface, pointing from "beneath" to "above" the surface. Then, the i'th component of the force exerted by the fluid below the surface on the fluid above the surface is

$$P_{i\alpha}\,\mathrm{d}S_\alpha\ ,$$

and the quantity of heat per unit time which flows from the fluid below across the surface to the fluid above is

$$\boldsymbol{q}\cdot\mathrm{d}\boldsymbol{S}\ .$$

3) For a discussion of the transport coefficients from a theoretical as well as experimental point of view, see J. O. Hirschfelder, C. F. Curtis and R. B. Bird, *Molecular Theory of Gases and Liquids* (John Wiley and Sons, Inc., New York, 1964). See esp. chapters 7, 8, 9.

4) That is, $\boldsymbol{u} = \boldsymbol{u}^{\perp} + \boldsymbol{u}^{\parallel}$, where curl $\boldsymbol{u}^{\parallel} = 0$ and div $\boldsymbol{u}^{\perp} = 0$.

5) To be specific, the dimensionless quantities:

$$\frac{k\kappa}{\rho v_0\, c_v}\ ,\quad \frac{k\eta}{\rho v_0}\ ,\quad \frac{k\zeta}{\rho v_0}\ ,$$

are assumed to be small. In most gases at 20°C and one atmosphere, for example, these quantities are of order unity when the wave length is about 10^{-5} cm.

6) Pogg. Ann. 134 (1868) 177. Or see *Gesammelte Abhandlungen*

von G.Kirchhoff (J.A.Barth, Leipzig, 1882) p.540. The Kirchhoff formula is usually expressed in terms of α, the absorption coefficient in amplitude for waves of a fixed frequency ω. To find α we must take ω real in (2.25) and solve for complex k. Then α is identified as the imaginary part of k. We find

$$\alpha = \frac{\omega^2}{2\rho v_0^3}\left[\tfrac{4}{3}\eta + \zeta + \kappa\left(\frac{1}{c_v} - \frac{1}{c_p}\right)\right] ,$$

which is the familiar result. The general result (2.23) was also obtained by Kirchhoff.

7) See, e.g. , S. Chapman and T. G. Cowling, *The Mathematical Theory of Non-Uniform Gases*, 2nd edition (Cambridge University Press, Cambridge, 1958) p. 241.

8) Other definitions of the mean free path differ from this by a numerical factor of order unity. See Chapman and Cowling, *op. cit.* chapter 5.

9) For the ideal gas, since the internal energy is independent of density, the relation (2.6) can be written:

$$T\,\mathrm{d}s = c_v\,\mathrm{d}T - \frac{p}{\rho^2}\,\mathrm{d}\rho .$$

Using this relation, the equation of state (2.28), the expression (2.13) for v_0, and the relations (2.21) and (2.29), we get the identities

$$\left(\frac{\partial\rho}{\partial s}\right)_p = \frac{\rho^2}{p}\left(\frac{c_v}{c_p} - 1\right) = -\frac{\rho}{c_p} ,$$

$$\left(\frac{\partial\rho}{\partial p}\right)_s = \frac{1}{v_0^2} ,$$

$$\frac{1}{T}\left(\frac{\partial T}{\partial p}\right)_s = \frac{p}{T c_v \rho^2}\left(\frac{\partial\rho}{\partial p}\right)_s = \frac{2}{3\rho v_0^2} ,$$

$$\frac{1}{T}\left(\frac{\partial T}{\partial s}\right)_p = \frac{1}{c_p} .$$

10) The authoratative reference is of course the book of Chapman and Cowling, *op. cit*. For a succinct summary see G. E. Uhlenbeck and G. W. Ford, *op.cit.*, chapter VI.

11) We restrict our consideration to the monoatomic gas, so the only coordinate is the position of the molecule. The case of polyatomic molecules is considered by C. S. Wang Chang, G. E. Uhlenbeck and J. de Boer, these *Studies*, Vol. II.

12) See Chapman and Cowling, *op. cit.*, p. 169.

13) See Chapman and Cowling, *op. cit.*, p. 174.

14) See Chapman and Cowling, *op. cit.*, p. 235.

15) See Chapman and Cowling, *op. cit.*, chapter 15.

16) H. Primakoff, J. Acoust. Soc. Am. 13 (1942) 14.
H. S. Tsien and R. Schamberg, J. Acoust. Soc. Am. 18 (1946) 334.

17) C. S. Wang Chang and G. E. Uhlenbeck, *On the Transport Phenomena in Rarefield Gases* (Johns Hopkins Applied Physics Laboratory Report, 1948) C. S. Wang Chang, *On the Dispersions of Sound in Helium* (Johns Hopkins Applied Physics Laboratory Report, 1948). These reports are reprinted with notes by G. E. Uhlenbeck in this *Study*.

18) C. S. Wang Chang, *op. cit.* For Helium molecules the force law used was that obtained by J. de Boer and A. Michels, Physica 5 (1938) 935; 6 (1939) 409.

Chapter III

THE LINEARIZED BOLTZMANN EQUATION

1. THE BOLTZMANN EQUATION

In kinetic theory an ideal monoatomic gas is completely described by the distribution function $f(\boldsymbol{r}, \boldsymbol{v}, t)$ where, to repeat, $f\,\mathrm{d}\boldsymbol{r}\,\mathrm{d}\boldsymbol{v}$ is the number of molecules in the element $\mathrm{d}\boldsymbol{r}\,\mathrm{d}\boldsymbol{v}$ of the six-dimensional phase space of a single particle. The time evolution of this distribution function is governed by the famous equation of Boltzmann [1]:

$$\frac{\partial f}{\partial t} + \boldsymbol{v} \cdot \frac{\partial f}{\partial \boldsymbol{r}} = \int \mathrm{d}\boldsymbol{v}_1 \int \mathrm{d}\Omega\, uI(u,\theta)\,[f'f_1' - ff_1] \;. \tag{3.1}$$

The left hand side of this equation is just the rate of change of f as seen by a freely moving molecule; the right hand side is the rate of change of f due to collisions between pairs of molecules. In the integrand the prime and the index 1 refer to velocity variables, e.g., $f_1' \equiv f(\boldsymbol{r}, \boldsymbol{v}_1', t)$, and the four velocities are those of a binary collision, in which a particle with velocity $\boldsymbol{v}$ collides with a particle with velocity $\boldsymbol{v}_1$, their velocities after collision being $\boldsymbol{v}'$ and $\boldsymbol{v}_1'$. In such a collision the velocity of the center of mass is unchanged;

$$\boldsymbol{U} = \tfrac{1}{2}(\boldsymbol{v} + \boldsymbol{v}_1) = \tfrac{1}{2}(\boldsymbol{v}' + \boldsymbol{v}_1') \;, \tag{3.2}$$

while the relative velocities before and after the collision,

$$\boldsymbol{u} = \boldsymbol{v} - \boldsymbol{v}_1 \;, \qquad \boldsymbol{u}' = \boldsymbol{v}' - \boldsymbol{v}_1' \;, \tag{3.3}$$

are equal in magnitude with an angle θ between them:

$$\boldsymbol{u} \cdot \boldsymbol{u}' = u^2 \cos\theta \;. \tag{3.4}$$

Finally, $I(u,\theta)$ is the collision cross section for a collision in which the relative velocity is rotated through an angle θ into the element of solid angle $\mathrm{d}\Omega = \sin\theta\,\mathrm{d}\theta\,\mathrm{d}\varphi$ [2].

The collision cross section is completely determined by the dynamics of a binary collision, but the question arises whether one should use classical or quantum mechanical dynamics [3]. The answer is that for all gases to which the Boltzmann equation applies, i.e., all gases whose equation of state is the classical ideal gas equation of state, the motion of the molecules in a binary collision is classical and we may use the classical cross section for all but very small angles. There are *always* quantum effects in small angle scattering, arising from the uncertainty principle, but they do not affect the Boltzmann equation since the factor $[f'f_1' - f f_1]$ vanishes at small angles [4]. However, we shall keep in mind the fact that these quantum effects have the result that the total cross section,

$$\sigma(u) = \int \mathrm{d}\Omega\, I(u,\theta) \;, \tag{3.5}$$

is in general finite.

Another quantum mechanical effect arises from the identity of the particles. This has the consequence that the cross section is in fact always symmetric around $\theta = \frac{1}{2}\pi$;

$$I(u,\theta) = I(u,\pi-\theta) \;, \tag{3.6}$$

although the classical cross section as it is usually defined (see note 2) does not have this symmetry. Again, this does not affect the Boltzmann equation, as we see if we interchange $\boldsymbol{v}'$ and $\boldsymbol{v}_1'$, the velocities of the particles after collision. Then $\boldsymbol{u}' \to -\boldsymbol{u}'$ and

$$\cos\theta = (\boldsymbol{u}\cdot\boldsymbol{u}')/u^2 \to -\cos\theta = \cos(\pi-\theta) \;. \tag{3.7}$$

Hence, (3.1) becomes

$$\frac{\partial f}{\partial t} + \boldsymbol{v}\cdot\frac{\partial f}{\partial \boldsymbol{r}} = \int \mathrm{d}\boldsymbol{v}_1 \int \mathrm{d}\Omega\, u I(u,\pi-\theta)\,[f'f_1' - f f_1] \;. \tag{3.8}$$

Adding equations (3.1) and (3.8) and dividing by 2, we can write the Boltzmann equation in the form:

$$\frac{\partial f}{\partial t} + \boldsymbol{v}\cdot\frac{\partial f}{\partial \boldsymbol{r}} = \int \mathrm{d}\boldsymbol{v} \int \mathrm{d}\Omega\, u\, I_{\mathrm{s}}(u,\theta)\,[f'f_1' - f f_1] \;, \tag{3.9}$$

where

$$I_{\mathrm{s}}(u,\theta) = \tfrac{1}{2}[I(u,\theta) + I(u,\pi-\theta)] \tag{3.10}$$

is the symmetric part of the collision cross section.

It is a well-known consequence of the Boltzmann equation that any initial distribution function approaches in time the Maxwell equilibrium distribution, which in the absence of external forces takes the form:

$$f_0(v) = n(m/2\pi kT)^{\frac{3}{2}} \, \mathrm{e}^{-mv^2/2kT} \,, \tag{3.11}$$

where n, the number density, and T, the absolute temperature, are constant throughout the gas. (Note that here k is Boltzmann's constant.) That this equilibrium distribution is indeed a solution of the Boltzmann equation follows from the fact that it is independent of position and time, so the left hand side of (3.1) vanishes, and that, since energy is conserved in a collision:

$$f_0(v)\, f_0(v_1) \equiv f_0(v')\, f_0(v_1') \,, \tag{3.12}$$

so the right hand side also vanishes.

2. THE LINEARIZED BOLTZMANN EQUATION

The motion of the gas which we call sound is a small amplitude disturbance from equilibrium. In seeking to describe such motions, it is appropriate to put in the Boltzmann equation:

$$f(\boldsymbol{r}, \boldsymbol{v}, t) = f_0(v)[1 + h(\boldsymbol{r}, \boldsymbol{v}, t)] \,, \tag{3.13}$$

and neglect quadratic terms in h [5]. Using the property (3.12) of the Maxwell distribution $f_0(v)$ we obtain the *linearized Boltzmann equation:*

$$\frac{\partial h}{\partial t} + \boldsymbol{v} \cdot \frac{\partial h}{\partial \boldsymbol{r}} = n\sigma \left(\frac{2kT}{m}\right)^{\frac{1}{2}} \boldsymbol{J}h \,, \tag{3.14}$$

where the *linearized collision operator,* $\boldsymbol{J}$, is given by

$$\boldsymbol{J}h = \frac{1}{n\sigma}\left(\frac{m}{2kT}\right)^{\frac{1}{2}} \int \mathrm{d}\boldsymbol{v}_1\, f_0(v_1) \int \mathrm{d}\Omega\, u\, I(u, \theta)[h' + h_1' - h - h_1] \,. \tag{3.15}$$

Here again, the prime and the index refer to the four velocities in a binary collision, $(\boldsymbol{v}, \boldsymbol{v}_1) \leftrightarrow (\boldsymbol{v}', \boldsymbol{v}_1')$. In these expressions σ is a constant with the dimensions of a cross section introduced so that $\boldsymbol{J}$ is dimensionless.

Our problem now is to determine the form of the solutions of this equation, to identify those special solutions we call sound, and to determine the dispersion law for these solutions. As a first step it is clearly of interest to investigate the properties of the linear operator $\boldsymbol{J}$.

It will prove convenient in the subsequent discussion to introduce dimensionless velocity variables:

$$\boldsymbol{c} = (m/2\,kT)^{\frac{1}{2}}\,\boldsymbol{v} \quad . \tag{3.16}$$

The linearized collision operator (3.15) may then be written:

$$\boldsymbol{J}h = \pi^{-\frac{3}{2}} \int \mathrm{d}\boldsymbol{c}_1\, \mathrm{e}^{-c_1^2} \int \mathrm{d}\Omega\, F(g,\theta)[h(\boldsymbol{c}') + h(\boldsymbol{c}_1') - h(\boldsymbol{c}) - h(\boldsymbol{c}_1)] \;, \tag{3.17}$$

where

$$F(g,\theta) = \frac{1}{\sigma} g\, I(u,\theta) \;, \tag{3.18}$$

with

$$\boldsymbol{g} = (m/2\,kT)^{\frac{1}{2}}\,\boldsymbol{u} \tag{3.19}$$

the dimensionless relative velocity. Note that $F(g,\theta)$ in general depends upon temperature since the cross section, when expressed in terms of the dimensionless relative velocity, becomes temperature dependent.

The linearized collision operator is a formally self-adjoint linear operator. To show this we first define the scalar product of two functions of $\boldsymbol{c}$, say $\chi(\boldsymbol{c})$ and $\varphi(\boldsymbol{c})$, to be

$$(\chi,\varphi) \equiv \pi^{-\frac{3}{2}} \int \mathrm{d}\boldsymbol{c}\, \mathrm{e}^{-c^2}\, \chi(\boldsymbol{c})\varphi(\boldsymbol{c}) \;. \tag{3.20}$$

Using straightforward manipulations we can show that [6]:

$$(\chi, \boldsymbol{J}\varphi) = -\frac{1}{4\pi^3} \int \mathrm{d}\boldsymbol{c} \int \mathrm{d}\boldsymbol{c}_1\, \mathrm{e}^{-c^2-c_1^2} \int \mathrm{d}\Omega\, F(g,\theta)$$

$$\times [\chi' + \chi_1' - \chi - \chi_1]\,[\varphi' + \varphi_1' - \varphi - \varphi_1] \;. \tag{3.21}$$

The fact that $\boldsymbol{J}$ is self-adjoint, i.e. that

$$(\chi, \boldsymbol{J}\varphi) = (\boldsymbol{J}\chi, \varphi) \;, \tag{3.22}$$

follows from the symmetry of the expression (3.21) with respect to interchange of χ and φ.

Consider next the eigenfunctions, ψ, and eigenvalues, λ, of $\boldsymbol{J}$:

$$\boldsymbol{J}\psi = \lambda\psi \ . \tag{3.23}$$

We see immediately that:

i) There are exactly five eigenfunctions with eigenvalue zero. These are

$$1 \ , \ \boldsymbol{c} \ , \ c^2 \ , \tag{3.24}$$

corresponding to the five additive constants of the motion in a binary collision: number of particles, the three components of momentum, and energy.

ii) All other eigenvalues are negative. To see this we note from (3.21) that

$$(\varphi, \boldsymbol{J}\varphi) = -\frac{1}{4\pi^3}\int \mathrm{d}\boldsymbol{c}\int \mathrm{d}\boldsymbol{c}_1 \, \mathrm{e}^{-c^2-c_1^2}\int \mathrm{d}\Omega F(g,\theta)[\varphi' + \varphi_1' - \varphi - \varphi_1]^2 \tag{3.25}$$

is less than or equal to zero for an arbitrary function $\varphi(\boldsymbol{c})$. Moreover, this quantity is equal to zero only if

$$\varphi' + \varphi_1' - \varphi - \varphi_1 = 0 \ ,$$

i.e. only if φ is a linear combination of the five additive constants of the motion. Hence, $\boldsymbol{J}$ is what is termed a negative semi-definite operator, whose spectrum of eigenvalues is non-positive [7].

iii) The operator $\boldsymbol{J}$ is a scalar operator. That is, the function $\boldsymbol{J}h$, given by (3.15), transforms under rotations in velocity space exactly as does the function $h(\boldsymbol{c})$. This has the consequence that the eigenfunctions of $\boldsymbol{J}$ must be of the form

$$\psi_{rlm}(\boldsymbol{c}) = R_{rl}(c)Y_{lm}(\hat{\boldsymbol{c}}) \ , \tag{3.26}$$

where $R_{rl}(c)$ is a function of the magnitude of $\boldsymbol{c}$ alone and $Y_{lm}(\hat{\boldsymbol{c}})$ is the spherical harmonic [8]. Moreover the eigenvalues λ_{rl} are independent of m and are, therefore, at least $2l+1$ fold degenerate. It should be emphasized that, although the labels l and m have only integer values, there is no reason to expect that the remaining label r will have only discrete values.

To see something of the physical significance of the eigenvalues of $\boldsymbol{J}$, consider the solution of the linearized Boltzmann equation for the *spatially homogeneous* case. If we assume we can develop the initial value of $h(\boldsymbol{c}, t)$ in terms of the eigenfunctions of $\boldsymbol{J}$;

$$h(\boldsymbol{c},0) = \sum_{r,l,m} a_{rlm}\psi_{rlm}(\boldsymbol{c}) \;, \tag{3.27}$$

then the solution of (3.14) is

$$h(\boldsymbol{c},t) = \sum_{r,l,m} \mathrm{e}^{-t/\tau_{rl}} \; a_{rlm}\psi_{rlm}(\boldsymbol{c}) \;, \tag{3.28}$$

where

$$(\tau_{rl})^{-1} = -\,n\sigma\left(\frac{2\,kT}{m}\right)^{\frac{1}{2}} \lambda_{rl} \,. \tag{3.29}$$

The fact that all the λ_{rl} are negative, excepting the five-fold degenerate eigenvalue zero belonging to the conserved quantities (3.24), means that an initial perturbation from equilibrium vanishes for long times except for permanent changes in density, average flow velocity, and internal energy (temperature). The τ_{rl} represent a spectrum of relaxation times for this approach to equilibrium.

3. THE HILBERT-ENSKOG CANONICAL FORM FOR $\boldsymbol{J}$

If the total scattering cross section is finite, which, as noted in section 1, is always the case if we correctly take into account the quantum effects in small angle scattering, we can express the collision operator in the form:

$$\boldsymbol{J}h = -\,m(c)h + \boldsymbol{K}h \;, \tag{3.30}$$

where

$$m(c) = \pi^{-\frac{3}{2}} \int \mathrm{d}\boldsymbol{c}_1\, \mathrm{e}^{-c_1^2} \int \mathrm{d}\Omega F(g,\theta) \;, \tag{3.31}$$

and where $\boldsymbol{K}$, which we call the Hilbert operator, is given by:

$$\boldsymbol{K}h = \pi^{-\frac{3}{2}} \int \mathrm{d}\boldsymbol{c}_1\, \mathrm{e}^{-c_1^2} \int \mathrm{d}\Omega F(g,\theta)[h(\boldsymbol{c}') + h(\boldsymbol{c}_1') - h(\boldsymbol{c}_1)] \,. \tag{3.32}$$

It was shown long ago by Enskog that $\boldsymbol{K}h$ can be written in the standard form of an integral operator [9]. That is, we can write

$$\boldsymbol{K}h = \pi^{-\frac{3}{2}} \int \mathrm{d}\boldsymbol{c}_1\, \mathrm{e}^{-c_1^2} K(\boldsymbol{c},\boldsymbol{c}_1)h(\boldsymbol{c}_1) \;, \tag{3.33}$$

where

$$K(\boldsymbol{c}, \boldsymbol{c}_1) = \int \mathrm{d}\Omega \left\{ [F(g \sec(\tfrac{1}{2}\theta), \theta) + F(g \sec(\tfrac{1}{2}\theta), \pi - \theta)] \sec^3(\tfrac{1}{2}\theta) \right.$$

$$\left. \times\, \mathrm{e}^{-g^2 \tan^2(\frac{1}{2}\theta)}\, I_0(2|\boldsymbol{c} \times \boldsymbol{c}_1| \tan(\tfrac{1}{2}\theta)) - F(g, \theta) \right\} . \tag{3.34}$$

Here $I_0(x)$ is the modified Bessel function with index zero [10].

To obtain this result we first remind ourselves that in (3.32) the velocity of the centre of mass is the same before and after collision,

$$\boldsymbol{G} = \tfrac{1}{2}(\boldsymbol{c} + \boldsymbol{c}_1) = \tfrac{1}{2}(\boldsymbol{c}' + \boldsymbol{c}_1') , \tag{3.35}$$

while the relative velocity after collision,

$$\boldsymbol{g}' = \boldsymbol{c}' - \boldsymbol{c}_1' , \tag{3.36}$$

is equal in magnitude to that before collision,

$$\boldsymbol{g} = \boldsymbol{c} - \boldsymbol{c}_1 , \tag{3.37}$$

but rotated through angle θ,

$$\cos\theta = \boldsymbol{g}\cdot\boldsymbol{g}'/g^2 . \tag{3.38}$$

The angular integral is over all directions of $\boldsymbol{g}'$. We may, therefore, rewrite (3.32) in the form:

$$\boldsymbol{K}h = \pi^{-\frac{3}{2}} \int \mathrm{d}\boldsymbol{c}_1\, \mathrm{e}^{-c_1^2} \int \mathrm{d}\boldsymbol{g}'\, \mathcal{F}(\boldsymbol{g}, \boldsymbol{g}') [2h(\tfrac{1}{2}(\boldsymbol{c} + \boldsymbol{c}_1 + \boldsymbol{g}')) - h(\boldsymbol{c}_1)] , \tag{3.39}$$

where

$$\mathcal{F}(\boldsymbol{g}, \boldsymbol{g}') = \frac{1}{2g^2} [F(g, \theta) + F(g, \pi - \theta)]\, \delta(g' - g) \tag{3.40}$$

with $\delta(g' - g)$ the Dirac delta function, and θ determined from (3.38). Note that we have defined $\mathcal{F}(\boldsymbol{g}, \boldsymbol{g}')$ to be symmetric about $\theta = \frac{1}{2}\pi$, so

$$\mathcal{F}(\boldsymbol{g}, -\boldsymbol{g}') = \mathcal{F}(\boldsymbol{g}, \boldsymbol{g}') . \tag{3.41}$$

This has enabled us to combine the first two terms in the square bracket of (3.32) into a single term in (3.39). If now we replace $\boldsymbol{c}_1$ with $2\boldsymbol{c}_1 - \boldsymbol{c} + \boldsymbol{g}'$, and note that $\mathrm{d}\boldsymbol{c}_1 \to 8\mathrm{d}\boldsymbol{c}_1$ and $\boldsymbol{g} \to 2\boldsymbol{g} - \boldsymbol{g}'$ under this change of variables, we can write Kh in the form (3.33), with

$$K(\boldsymbol{c},\boldsymbol{c}_1) = \int \mathrm{d}\boldsymbol{g}' \left[16\mathcal{F}(2\boldsymbol{g}-\boldsymbol{g}',\boldsymbol{g}')\, \mathrm{e}^{c_1^2-(\boldsymbol{c}_1-\boldsymbol{g}+\boldsymbol{g}')^2} - \mathcal{F}(\boldsymbol{g},\boldsymbol{g}')\right] . \tag{3.42}$$

Here the last term is

$$\int \mathrm{d}\boldsymbol{g}'\, \mathcal{F}(\boldsymbol{g},\boldsymbol{g}') = \int \mathrm{d}\Omega \int_0^\infty \mathrm{d}g'\, F(g,\theta)\,\delta(g'-g) = \int \mathrm{d}\Omega\, F(g,\theta) . \tag{3.43}$$

The first term in (3.42) is

$$16 \int \mathrm{d}\boldsymbol{g}'\, \mathcal{F}(2\boldsymbol{g}-\boldsymbol{g}',\boldsymbol{g}')\, \mathrm{e}^{c_1^2-(\boldsymbol{c}_1-\boldsymbol{g}+\boldsymbol{g}')^2} \tag{3.44}$$

$$= 16 \int \mathrm{d}\Omega' \int_0^\infty \mathrm{d}g'\, F(g',\theta)\,\delta(g'-|2\boldsymbol{g}-\boldsymbol{g}'|)\,\exp\{-|\boldsymbol{g}-\boldsymbol{g}'|^2 + 2\boldsymbol{c}_1\cdot(\boldsymbol{g}-\boldsymbol{g}')\} ,$$

where $\mathrm{d}\Omega' = \sin\psi\,\mathrm{d}\psi\,\mathrm{d}\varphi$ with ψ the angle between $\boldsymbol{g}'$ and $\boldsymbol{g}$. The delta function requires that $\boldsymbol{g}'$ and $2\boldsymbol{g} - \boldsymbol{g}'$ form two sides of an equilateral triangle whose base is $2\boldsymbol{g}$ (see fig. 3). It follows that

$$g' = g \sec\psi = g \sec(\tfrac{1}{2}\theta) , \tag{3.45}$$

and

$$\boldsymbol{g} - \boldsymbol{g}' = \tan(\tfrac{1}{2}\theta)\,\hat{\boldsymbol{n}} \times \boldsymbol{g} , \tag{3.46}$$

where $\hat{\boldsymbol{n}}$ is the unit normal to the plane of $\boldsymbol{g}$ and $\boldsymbol{g}'$. The delta function is [11]

$$\delta(g' - |2g - g'|) = \tfrac{1}{2} \sec^2(\tfrac{1}{2}\theta)\,\delta(g' - g\sec(\tfrac{1}{2}\theta)) . \tag{3.47}$$

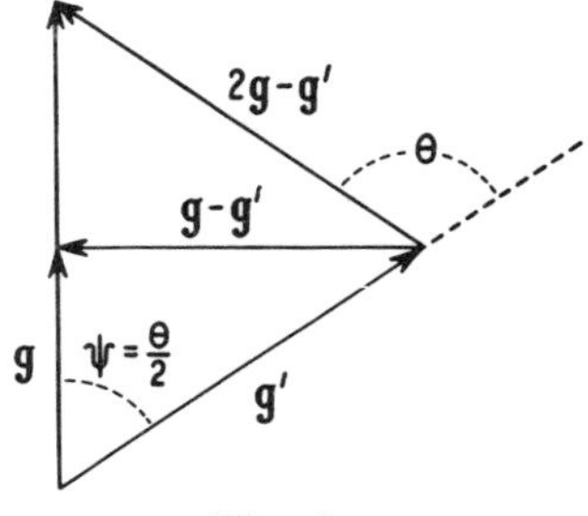

Fig. 3.

Using the fact that

$$d\Omega' = \sin\psi \, d\psi d\varphi = \tfrac{1}{4} \sec(\tfrac{1}{2}\theta) \sin\theta \, d\theta \, d\varphi = \tfrac{1}{4} \sec(\tfrac{1}{2}\theta) \, d\Omega \,, \tag{3.48}$$

we find (3.44) becomes

$$\int d\Omega \sec^3(\tfrac{1}{2}\theta)[\boldsymbol{F}(g\sec(\tfrac{1}{2}\theta), \theta) + \boldsymbol{F}(g\sec(\tfrac{1}{2}\theta), \pi-\theta)] \times \exp\{-g^2\tan^2(\tfrac{1}{2}\theta) + 2\tan(\tfrac{1}{2}\theta)\hat{\boldsymbol{n}}\cdot(\boldsymbol{c}\times\boldsymbol{c}_1)\} \,. \tag{3.49}$$

But $\hat{\boldsymbol{n}}\cdot(\boldsymbol{c}\times\boldsymbol{c}_1) = |\boldsymbol{c}\times\boldsymbol{c}_1| \cos\varphi$, and

$$I_0(x) = \frac{1}{2\pi}\int_0^{2\pi} d\varphi \, e^{x\cos\varphi} \,, \tag{3.50}$$

so (3.49) becomes

$$2\pi \int_0^{\pi} d\theta \sin\theta \sec^3(\tfrac{1}{2}\theta)[\boldsymbol{F}(g\sec(\tfrac{1}{2}\theta), \theta) + \boldsymbol{F}(g\sec(\tfrac{1}{2}\theta), \pi-\theta)] \times e^{-g^2\tan^2(\frac{1}{2}\theta)} I_0(2|\boldsymbol{c}\times\boldsymbol{c}_1|\tan(\tfrac{1}{2}\theta)) \,. \tag{3.51}$$

Since this expression is independent of the azimuthal angle φ, it is unchanged if we integrate over φ from 0 to 2π and then divide by 2π. Hence, we see that, combining (3.43) and (3.51), we obtain (3.34).

For the special case of elastic spheres of diameter a,

$$F(g, \theta) = \frac{a^2}{4\sigma} g \,. \tag{3.52}$$

From (3.31) we have, therefore,

$$m(\boldsymbol{c}) = \frac{\pi a^2}{\sigma} \pi^{-\frac{3}{2}} \int d\boldsymbol{c}_1 \, e^{-c_1^2} |\boldsymbol{c}-\boldsymbol{c}_1| = \frac{\pi a^2}{\sigma} \pi^{-\frac{1}{2}} \left[e^{-c^2} + \left(2c + \frac{1}{c}\right) \int_0^c du \, e^{-u^2} \right] . \tag{3.53}$$

From (3.34) we get Hilbert's expression:

$$K(\boldsymbol{c}, \boldsymbol{c}_1) = \frac{\pi a^2}{\sigma} \left(\frac{2}{g} e^{|\boldsymbol{c}\times\boldsymbol{c}_1|^2/g^2} - g \right) . \tag{3.54}$$

For this case of elastic spheres an obvious choice of σ is the total cross section, which would make the factor $\pi a^2/\sigma$ equal to unity in these two equations.

4. THE MATRIX ELEMENTS WITH RESPECT TO THE BURNETT FUNCTIONS

The matrix elements of the linearized collision operator with respect to the so-called Burnett functions have been of great use for practical calculations [12]. These functions form a complete set of orthogonal functions in velocity space and are given by:

$$\chi_{rlm}(\boldsymbol{c}) = c^l\, L_r^{(l+\frac{1}{2})}(c^2)\, Y_{lm}(\hat{\boldsymbol{c}}) \;, \tag{3.55}$$

where $Y_{lm}(\hat{\boldsymbol{c}})$ is the spherical harmonic and

$$L_r^{(\alpha)}(z) = \frac{1}{r!}\, z^{-\alpha}\, e^z\, \frac{d^r}{dz^r}\, e^{-z}\, z^{r+\alpha} \tag{3.56}$$

is the associated Laguerre polynomial [13]. The functions χ_{rlm} are orthogonal but not normalized. In fact

$$(\chi^*_{rlm},\, \chi_{r'l'm'}) = \frac{\Gamma(r+l+\frac{3}{2})}{4\pi\, r!\, \Gamma(\frac{3}{2})}\, \delta_{rr'}\, \delta_{ll'}\, \delta_{mm'} \;. \tag{3.57}$$

We denote the normalized Burnett functions by ψ_{rlm}:

$$\psi_{rlm}(\boldsymbol{c}) = \left[\frac{4\pi r!\, \Gamma(\frac{3}{2})}{\Gamma(r+l+\frac{3}{2})}\right]^{\frac{1}{2}} \chi_{rlm}(\boldsymbol{c}) \;. \tag{3.58}$$

Since the linearized collision operator, $\boldsymbol{J}$, is a scalar operator, its matrix elements with respect to the Burnett functions must be diagonal in l and m, and independent of m. Thus

$$(\psi^*_{rlm},\, \boldsymbol{J}\psi_{r'l'm'}) = J^l_{rr'}\, \delta_{ll'}\, \delta_{mm'} \;. \tag{3.59}$$

It will be convenient to consider the matrix elements of $\boldsymbol{J}$ with respect to the unnormalized Burnett functions:

$$(\chi^*_{rlm},\, \boldsymbol{J}\chi_{r'l'm'}) = M^l_{rr'}\, \delta_{ll'}\, \delta_{mm'} \;, \tag{3.60}$$

where

$$J^l_{rr'} = 4\pi\, \Gamma(\tfrac{3}{2}) \left[\frac{r!\, r'!}{\Gamma(r+l+\frac{3}{2})\, \Gamma(r'+l+\frac{3}{2})}\right]^{\frac{1}{2}} M^l_{rr'} \;. \tag{3.61}$$

From (3.60) we see we can write

$$M^l_{rr'} = \frac{1}{2l+1} \sum_{m=-l}^{l} (\chi^*_{rlm},\, \boldsymbol{J}\chi_{r'lm}) \;. \tag{3.62}$$

Inserting the form (3.17) for $\boldsymbol{J}$, we obtain

$$M^l_{rr'} = \frac{1}{4\pi^4} \int \mathrm{d}c \int \mathrm{d}c_1\, \mathrm{e}^{-c^2 - c_1^2}\, c^l\, L_r^{(l+\frac{1}{2})}(c^2) \int \mathrm{d}\Omega\, F(g,\theta)$$

$$\times \Big[c'^l\, L_{r'}^{(l+\frac{1}{2})}(c'^2)\, P_l(\hat{c}\cdot\hat{c}') + c_1'^l\, L_{r'}^{(l+\frac{1}{2})}(c_1'^2)\, P_l(\hat{c}\cdot\hat{c}_1')$$

$$- c^l\, L_{r'}^{(l+\frac{1}{2})}(c^2) - c_1^l\, L_{r'}^{(l+\frac{1}{2})}(c_1^2) P_l(\hat{c}\cdot\hat{c}_1) \Big] . \qquad (3.63)$$

Here we have used the addition theorem for spherical harmonics [14]:

$$\sum_{m=-l}^{l} Y^*_{lm}(\hat{c})\, Y_{lm}(\hat{c}_1) = \frac{2l+1}{4\pi} P_l(\hat{c}\cdot\hat{c}_1) . \qquad (3.64)$$

Next we note the generating function for the Laguerre polynomials [13]:

$$(1-x)^{-\alpha-1}\, \mathrm{e}^{-xz/(1-x)} = \sum_{r=0}^{\infty} L_r^{(\alpha)}(z)\, x^r . \qquad (3.65)$$

Hence, the matrix elements (3.63) may be obtained from the generating function:

$$M^l(x,y) \equiv \sum_{r=0}^{\infty} \sum_{r'=0}^{\infty} M^l_{rr'}\, x^r y^{r'}$$

$$= \frac{1}{4\pi^4[(1-x)(1-y)]^{l+\frac{3}{2}}} \int \mathrm{d}c \int \mathrm{d}c_1\, \mathrm{e}^{-c^2/(1-x) - c_1^2} \int \mathrm{d}\Omega\, F(g,\theta)$$

$$\times \Big[\mathrm{e}^{-yc'^2/(1-y)} (cc')^l\, P_l(\hat{c}\cdot\hat{c}') + \mathrm{e}^{-yc_1'^2/(1-y)} (cc_1')^l\, P_l(\hat{c}\cdot\hat{c}_1')$$

$$- c^{2l}\, \mathrm{e}^{-yc^2/(1-y)} - \mathrm{e}^{-yc_1^2/(1-y)} (cc_1)^l\, P_l(\hat{c}\cdot\hat{c}_1) \Big] . \qquad (3.66)$$

We use next the identity [15]:

$$\sum_{l=0}^{\infty} (cc_1)^l\, P_l(\hat{c}\cdot\hat{c}_1) \frac{t^l}{l!} = \frac{1}{2\pi} \int_0^{2\pi} \mathrm{d}\alpha\, \mathrm{e}^{t(c\,\cdot\, c_1) + it\,\hat{l}\cdot(c\times c_1)} , \qquad (3.67)$$

where the integral is over all directions of the unit vector $\hat{l}$ which

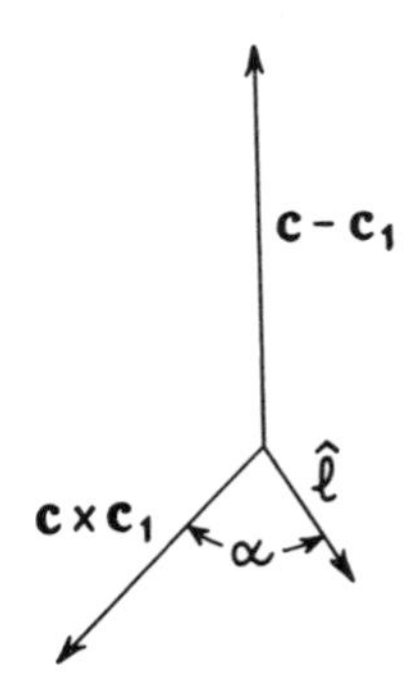

Fig. 4. The unit vector $\hat{l}$ lies in the plane perpendicular to $c - c_1$.

are perpendicular to $c - c_1$. (See fig. 4). Hence, if we form the generating function:

$$M(x,y;t) = \sum_{l=0}^{\infty} M^l(x,y) \frac{t^l}{l!} , \tag{3.68}$$

then we can write:

$$M(x,y;t) = \frac{1}{8\pi^5[(1-x)(1-y)]^{\frac{3}{2}}} \int \mathrm{d}c \int \mathrm{d}c_1 \int \mathrm{d}\Omega\, F(g,\theta) \int_0^{2\pi} \mathrm{d}\alpha$$

$$\times \left[\exp\left\{ -\frac{c^2}{1-x} - c_1^2 - \frac{yc'^2}{1-y} + \frac{t(c\cdot c' + i\hat{l}\cdot(c\times c'))}{(1-x)(1-y)} \right\} \right.$$

$$+ \exp\left\{ -\frac{c^2}{1-x} - c_1^2 - \frac{yc_1'^2}{1-y} + \frac{t(c\cdot c_1' + i\hat{l}\cdot(c\times c_1'))}{(1-x)(1-y)} \right\}$$

$$- \exp\left\{ -\frac{(1-xy-t)c^2}{(1-x)(1-y)} - c_1^2 \right\}$$

$$\left. - \exp\left\{ -\frac{c^2}{1-x} - \frac{c_1^2}{1-y} + \frac{t(c\cdot c_1 + i\hat{l}\cdot(c\times c_1))}{(1-x)(1-y)} \right\} \right] . \tag{3.69}$$

Here the α integration is over all directions of $\hat{l}$ which are perpendicular to $\boldsymbol{c}-\boldsymbol{c}'$ in the first term, $\boldsymbol{c}-\boldsymbol{c}_1'$ in the second term, and $\boldsymbol{c}-\boldsymbol{c}_1$ in the last term in square brackets. The third term is independent of α so the α integration gives just a factor of 2π.

We now change variables to $\boldsymbol{G}$ and $\boldsymbol{g}$ given by (3.35) and (3.37). That is, we put

$$\begin{aligned} \boldsymbol{c} &= \boldsymbol{G} + \tfrac{1}{2}\boldsymbol{g} , \\ \boldsymbol{c}_1 &= \boldsymbol{G} - \tfrac{1}{2}\boldsymbol{g} , \end{aligned} \tag{3.70}$$

while

$$\begin{aligned} \boldsymbol{c}' &= \boldsymbol{G} + \tfrac{1}{2}\boldsymbol{g}' , \\ \boldsymbol{c}_1' &= \boldsymbol{G} - \tfrac{1}{2}\boldsymbol{g}' , \end{aligned} \tag{3.71}$$

with $|\boldsymbol{g}'| = |\boldsymbol{g}| = g$. Since the transformation (3.70) has unit Jacobean, $\mathrm{d}\boldsymbol{c}\,\mathrm{d}\boldsymbol{c}_1 = \mathrm{d}\boldsymbol{G}\,\mathrm{d}\boldsymbol{g}$, and equation (3.69) becomes:

$$\begin{aligned} M(x,y;t) = {} & \frac{1}{8\pi^5[(1-x)(1-y)]^{\frac{3}{2}}} \int \mathrm{d}\boldsymbol{G} \int \mathrm{d}\boldsymbol{g} \int \mathrm{d}\Omega\, F(g,\theta) \exp\left\{-\frac{2-x-y-t}{(1-x)(1-y)} G^2\right\} \\ & \times \int_0^{2\pi} \mathrm{d}\alpha \left[\exp\left\{-\frac{x(1-y)\boldsymbol{g}+y(1-x)\boldsymbol{g}'-\frac{1}{2}t(\boldsymbol{g}+\boldsymbol{g}')-\frac{1}{2}it\hat{\boldsymbol{l}}\times(\boldsymbol{g}-\boldsymbol{g}')}{(1-x)(1-y)} \cdot \boldsymbol{G} \right.\right. \\ & \left. - \frac{(2-x-y)g^2 - t(\boldsymbol{g}\cdot\boldsymbol{g}' + i\hat{\boldsymbol{l}}\cdot(\boldsymbol{g}\times\boldsymbol{g}'))}{4(1-x)(1-y)} \right\} \\ & + \exp\left\{-\frac{x(1-y)\boldsymbol{g}-y(1-x)\boldsymbol{g}'-\frac{1}{2}t(\boldsymbol{g}-\boldsymbol{g}')-\frac{1}{2}it\hat{\boldsymbol{l}}\times(\boldsymbol{g}+\boldsymbol{g}')}{(1-x)(1-y)} \cdot \boldsymbol{G} \right. \\ & \left. - \frac{(2-x-y)g^2 + t(\boldsymbol{g}\cdot\boldsymbol{g}' + i\hat{\boldsymbol{l}}\cdot(\boldsymbol{g}\times\boldsymbol{g}'))}{4(1-x)(1-y)} \right\} \\ & - \exp\left\{-\frac{x(1-y)+y(1-x)-t}{(1-x)(1-y)}\, \boldsymbol{g}\cdot\boldsymbol{G} - \frac{2-x-y-t}{4(1-x)(1-y)}\, g^2 \right\} \\ & \left. - \exp\left\{-\frac{(x-y)\boldsymbol{g}-it\hat{\boldsymbol{l}}\times\boldsymbol{g}}{(1-x)(1-y)} \cdot \boldsymbol{G} - \frac{2-x-y+t}{4(1-x)(1-y)}\, g^2 \right\} \right] . \end{aligned} \tag{3.72}$$

Here the α integration is over all directions of $\hat{\boldsymbol{l}}$ perpendicular to $\boldsymbol{g}-\boldsymbol{g}'$ in the first term, $\boldsymbol{g}+\boldsymbol{g}'$ in the second term, and $\boldsymbol{g}$ in the last term. We now perform the $\boldsymbol{G}$ integration, using the integral formula:

$$\int \mathrm{d}\boldsymbol{G}\, \mathrm{e}^{-AG^2+\boldsymbol{B}\cdot\boldsymbol{G}} = \left(\frac{\pi}{A}\right)^{\frac{3}{2}} \exp\frac{B^2}{4A}\,. \tag{3.73}$$

We find

$$M(x,y;t) = \frac{1}{4\pi^{\frac{5}{2}}(2-x-y-t)^{\frac{3}{2}}} \int \mathrm{d}\boldsymbol{g} \exp\left\{-\frac{2-xy-t}{2(2-x-y-t)}g^2\right\} \int \mathrm{d}\Omega\, F(g,\theta)$$

$$\times\left[\frac{1}{2\pi}\int_0^{2\pi} \mathrm{d}\alpha \exp\left\{\frac{(xy+t)\boldsymbol{g}\cdot\boldsymbol{g}' + it\hat{\boldsymbol{l}}\cdot(\boldsymbol{g}\times\boldsymbol{g}')}{2(2-x-y-t)}\right\}\right.$$

$$+\frac{1}{2\pi}\int_0^{2\pi} \mathrm{d}\alpha \exp\left\{-\frac{(xy+t)\boldsymbol{g}\cdot\boldsymbol{g}' + it\hat{\boldsymbol{l}}\cdot(\boldsymbol{g}\times\boldsymbol{g}')}{2(2-x-y-t)}\right\}$$

$$\left.-\,2\cosh\left\{\frac{(xy+t)g^2}{2(2-x-y-t)}\right\}\right]. \tag{3.74}$$

In each case the α integration is over all directions of $\hat{\boldsymbol{l}}$ which lie in a plane containing $\boldsymbol{g}\times\boldsymbol{g}'$. We may therefore write

$$\hat{\boldsymbol{l}}\cdot(\boldsymbol{g}\times\boldsymbol{g}') = |\boldsymbol{g}\times\boldsymbol{g}'|\cos\alpha = g^2\sin\theta\cos\alpha\,. \tag{3.75}$$

If we note the integral representation of Bessel's function of order zero [16]:

$$\mathcal{J}_0(z) = \frac{1}{2\pi}\int_0^{2\pi} \mathrm{d}\alpha\, \mathrm{e}^{iz\cos\alpha}\,, \tag{3.76}$$

we can write

$$M(x,y;t) = \frac{1}{2\pi^{\frac{5}{2}}(2-x-y-t)^{\frac{3}{2}}} \int \mathrm{d}\boldsymbol{g} \exp\left\{-\frac{2-xy-t}{2(2-x-y-t)}g^2\right\} \int \mathrm{d}\Omega\, F(g,\theta)$$

$$\times\left[\cosh\left\{\frac{(xy+t)g^2\cos\theta}{2(2-x-y-t)}\right\}\mathcal{J}_0\left\{\frac{tg^2\sin\theta}{2(2-x-y-t)}\right\}\right.$$

$$\left.-\cosh\left\{\frac{(xy+t)g^2}{2(2-x-y-t)}\right\}\right]. \tag{3.77}$$

This is as far as we can proceed without specifying the form of $F(g,\theta)$.

For general $F(g,\theta)$ we can expand (3.77) in powers of x, y and t and identify the matrix elements $M^{l}_{rr'}$ from the expansion:

$$M(x,y;t) = \sum_{l=0}^{\infty} \sum_{r=0}^{\infty} \sum_{r'=0}^{\infty} M^{l}_{rr'}\, x^{r} y^{r'} \frac{t^{l}}{l!} . \qquad (3.78)$$

Following Chapman and Cowling, we express the matrix elements in terms of the integrals [17]:

$$\Omega^{(j)}(k) = 2^{-k-\frac{7}{2}} \pi^{-\frac{1}{2}} \int \mathrm{d}\boldsymbol{g}\, \mathrm{e}^{-\frac{1}{2}g^{2}} g^{2k} \phi^{(j)}(g) , \qquad (3.79)$$

where

$$\phi^{(j)}(g) = \frac{\sigma}{2\pi}\left(\frac{2\,kT}{m}\right)^{\frac{1}{2}} \int \mathrm{d}\Omega\, F(g,\theta)(1-\cos^{j}\theta) . \qquad (3.80)$$

In appendix A we give expressions for the first few matrix elements obtained in this way.

For the case of repulsive power law potentials we can proceed farther. If the intermolecular potential is of the form:

$$\varphi(r) = \varphi_{s}\left(\frac{a}{r}\right)^{s} , \qquad (3.81)$$

then we can show in general that [2]

$$F(g,\theta) = \frac{a^{2}}{\sigma}\left(\frac{2\,\varphi_{s}}{kT}\right)^{\frac{2}{s}} g^{1-\frac{4}{s}} f_{s}(\theta) , \qquad (3.82)$$

where $f_{s}(\theta)$ is a purely numerical function of θ alone. Using this expression for $F(g,\theta)$, the expression (3.74) for $M(x,y;t)$ can be put in the form:

$$M(x,y;t) = \frac{a^{2}}{\sigma}\left(\frac{2\varphi_{s}}{kT}\right)^{\frac{1}{2}} \frac{1}{4\pi^{\frac{5}{2}}(2-x-y-t)^{\frac{3}{2}}} \int \mathrm{d}\Omega\, f_{s}(\theta)$$

$$\times \int \mathrm{d}\boldsymbol{g}\, g^{1-\frac{4}{s}} \Bigg[\frac{1}{2\pi}\int_{0}^{2\pi} \mathrm{d}\alpha\, \exp\left\{-\frac{2-xy-t-(xy+t)\cos\theta - it\sin\theta\cos\alpha}{2(2-x-y-t)}\, g^{2}\right\}$$

$$+ \frac{1}{2\pi}\int_{0}^{2\pi} \mathrm{d}\alpha\, \exp\left\{-\frac{2-xy-t+(xy+t)\cos\theta + it\sin\theta\cos\alpha}{2(2-x-y-t)}\, g^{2}\right\}$$

$$- \exp\left\{-\frac{1-xy-t}{2-x-y-t}\, g^{2}\right\} - \exp\left\{-\frac{1}{2-x-y-t} g^{2}\right\}\Bigg] . \qquad (3.83)$$

We now use the integral formula:

$$\pi^{-\frac{3}{2}} \int d\boldsymbol{g}\, g^{1-\frac{4}{s}}\, e^{-Ag^2} = \frac{\Gamma(2-\frac{2}{s})}{\Gamma(\frac{3}{2})} A^{\frac{2}{s}-2} , \qquad (3.84)$$

and then write $M(x,y;t)$ in the form:

$$M(x,y;t) = \frac{a^2}{\sigma}\left(\frac{2\varphi_s}{kT}\right)^{\frac{2}{s}} \frac{\Gamma(2-\frac{2}{s})}{4\pi\,\Gamma(\frac{3}{2})} (2-x-y-t)^{\frac{1}{2}-\frac{2}{s}}$$

$$\times \int d\Omega\, f_s(\theta)\Bigg[\frac{1}{2\pi}\int_0^{2\pi} d\alpha \Big\{1-xy\cos^2(\tfrac{1}{2}\theta) - t\cos(\tfrac{1}{2}\theta)\Big(\cos(\tfrac{1}{2}\theta) + i\sin(\tfrac{1}{2}\theta)\cos\alpha\Big)\Big\}^{\frac{2}{s}-2} .$$

$$+\frac{1}{2\pi}\int_0^{2\pi} d\alpha \Big\{1-xy\sin^2(\tfrac{1}{2}\theta) - t\sin(\tfrac{1}{2}\theta)\Big(\sin(\tfrac{1}{2}\theta) - i\cos(\tfrac{1}{2}\theta)\cos\alpha\Big)\Big\}^{\frac{2}{s}-2}$$

$$-(1-xy-t)^{\frac{2}{s}-2} - 1\Bigg] . \qquad (3.85)$$

Using the formula:

$$(1-A-B)^{-\nu} = \sum_{j=0}^{\infty}\sum_{k=0}^{\infty} \frac{\Gamma(\nu+j+k)}{\Gamma(\nu)j!\,k!} A^j B^k \qquad (3.86)$$

we can write:

$$\frac{1}{2\pi}\int_0^{2\pi} d\alpha \Big\{1-xy\cos^2(\tfrac{1}{2}\theta) - t\cos(\tfrac{1}{2}\theta)\Big(\cos(\tfrac{1}{2}\theta) + i\sin(\tfrac{1}{2}\theta)\cos\alpha\Big)\Big\}^{\frac{2}{s}-2}$$

$$= \sum_{j=0}^{\infty}\sum_{k=0}^{\infty} \frac{\Gamma(2-\frac{2}{s}+j+k)}{\Gamma(2-\frac{2}{s})j!\,k!} \cos^{2j+k}(\tfrac{1}{2}\theta)(xy)^j t^k \frac{1}{2\pi}\int_0^{2\pi} d\alpha \Big(\cos(\tfrac{1}{2}\theta) + i\sin(\tfrac{1}{2}\theta)\cos\alpha\Big)^k$$

$$= \sum_{j=0}^{\infty}\sum_{k=0}^{\infty} \frac{\Gamma(2-\frac{2}{s}+j+k)}{\Gamma(2-\frac{2}{s})j!\,k!} \cos^{2j+k}(\tfrac{1}{2}\theta)\, P_k(\cos(\tfrac{1}{2}\theta))(xy)^j t^k , \qquad (3.87)$$

where we have used the well-known integral representation of the Legendre polynomials 15). Inserting this and similar expressions for the other terms in (3.85), we can write

$$M(x,y;t) = \frac{a^2}{2^{\frac{1}{2}}\sigma}\left(\frac{\varphi_s}{kT}\right)^{\frac{2}{s}} \frac{\Gamma(2-\frac{2}{s})}{\Gamma(\frac{3}{2})}\left(1-\frac{x+y+t}{2}\right)^{\frac{1}{2}-\frac{2}{s}} \sum_{j=0}^{\infty}\sum_{k=0}^{\infty} B_k^j(s)(xy)^j(\tfrac{1}{2}t)^k \tag{3.88}$$

where

$$B_k^j(s) = \frac{2^{k+1}\Gamma(2-\frac{2}{s}+j+k)}{\Gamma(2-\frac{2}{s})\,j!\,k!}\frac{1}{4\pi}\int \mathrm{d}\Omega\, f_s(\theta)$$
$$\times\left[\cos^{2j+k}(\tfrac{1}{2}\theta)\,P_k(\cos(\tfrac{1}{2}\theta)) + \sin^{2j+k}(\tfrac{1}{2}\theta)\,P_k(\sin(\tfrac{1}{2}\theta)) - 1 - \delta_{j,0}\,\delta_{k,0}\right]. \tag{3.89}$$

If in (3.88) we expand:

$$\left(1-\frac{x+y+t}{2}\right)^{\frac{1}{2}-\frac{2}{s}} = \sum_{u=0}^{\infty}\sum_{v=0}^{\infty}\sum_{w=0}^{\infty}\frac{\Gamma(\frac{2}{s}-\frac{1}{2}+u+v+w)}{\Gamma(\frac{2}{s}-\frac{1}{2})\,u!\,v!\,w!}\,x^u y^v t^w , \tag{3.90}$$

we can write

$$M(x,y;t) = \frac{a^2}{2^{\frac{1}{2}}\sigma}\left(\frac{\varphi_s}{kT}\right)^{\frac{2}{s}}\frac{\Gamma(2-\frac{2}{s})}{\Gamma(\frac{3}{2})}$$
$$\times\sum_{u=0}^{\infty}\sum_{v=0}^{\infty}\sum_{w=0}^{\infty}\sum_{j=0}^{\infty}\sum_{k=0}^{\infty}\frac{2^{-u-v-w-k}\,\Gamma(\frac{2}{s}-\frac{1}{2}+u+v+w)}{\Gamma(\frac{2}{s}-\frac{1}{2})\,u!\,v!\,w!}B_k^j(s)$$
$$\times\; x^{u+j}\,y^{v+j}\,t^{w+k} . \tag{3.91}$$

If in these summations we replace

$$u = r-j\,, \qquad v = r'-j\,, \qquad w = l-k\,, \tag{3.92}$$

we can write (3.91) in the form (3.78) with

$$M_{rr'}^{l} = \frac{a^2}{\sigma}\left(\frac{\varphi_s}{kT}\right)^{\frac{2}{s}}\frac{\Gamma(2-\frac{2}{s})\,l!}{\Gamma(\frac{3}{2})\,2^{r+r'+l+\frac{1}{2}}}$$

$$\times\sum_{j=0}^{r,r'}\sum_{k=0}^{l}\frac{4^j\,\Gamma(\frac{2}{s}-\frac{1}{2}+r+r'-2j+l-k)}{\Gamma(\frac{2}{s}-\frac{1}{2})\,(r-j)!\,(r'-j)!\,(l-k)!}B_k^j(s). \tag{3.93}$$

Here the upper limit of the j summation is the smaller of r and r'.

The expression (3.93) is our desired expression for the matrix elements of $\boldsymbol{J}$ with respect to the unnormalized Burnett functions. The matrix elements with respect to the normalized Burnett functions are given by (3.61).

Of special interest are the cases of Maxwell molecules ($s=4$) and elastic spheres of diameter a ($s=\infty$). For the case of Maxwell molecules the Burnett functions are eigenfunctions of the collision operator; the matrix is diagonal with the diagonal elements the eigenvalues [18]. Thus

$$J^{l}_{rr'} = \lambda_{rl}\,\delta_{r,r'} \,, \qquad (s=4) \,, \tag{3.94}$$

with

$$\lambda_{rl} = \frac{a^2}{\sigma}\left(\frac{\varphi_4}{kT}\right)^{\frac{1}{2}} \frac{4\pi\,\Gamma(\frac{3}{2})r!\,l!}{2^{l+\frac{1}{2}}\,\Gamma(r+l+\frac{3}{2})} B^{r}_{l}(4)$$

$$= \frac{a^2}{\sigma}\left(\frac{2\varphi_4}{kT}\right)^{\frac{1}{2}} \int \mathrm{d}\Omega\; f_4(\theta)\Big[\cos^{2r+l}(\tfrac{1}{2}\theta)\,P_l(\cos(\tfrac{1}{2}\theta))$$

$$+ \sin^{2r+l}(\tfrac{1}{2}\theta)\,P_l(\sin(\tfrac{1}{2}\theta)) - 1 - \delta_{r,0}\,\delta_{l,0}\Big] \,. \tag{3.95}$$

These eigenvalues are all discrete, although there are some "accidental" degeneracies. The largest non-zero eigenvalue is

$$\lambda_{11} = \lambda_{20} = -\,1.94\,\frac{a^2}{\sigma}\left(\frac{2\varphi_4}{kT}\right)^{\frac{1}{2}} \,. \tag{3.96}$$

There is no lower bound on the eigenvalues [19].

We obtain the case of elastic spheres of diameter a if we let $s \to \infty$ in (3.93). Since for elastic spheres $I(u,\theta) = \frac{1}{4}a^2$, we see that $f_\infty(\theta) = \frac{1}{4}$. Hence, from (3.89) we have

$$B^{j}_{k}(\infty) = \frac{2^{k-2}(j+k+1)!}{j!\,k!} \int_0^{\pi} \mathrm{d}\theta\,\sin\theta\Big[\cos^{2j+k}(\tfrac{1}{2}\theta)\,P_k(\cos(\tfrac{1}{2}\theta))$$

$$+ \sin^{2j+k}(\tfrac{1}{2}\theta)\,P_k(\sin(\tfrac{1}{2}\theta)) - 1 - \delta_{j,0}\,\delta_{k,0}\Big] \,. \tag{3.97}$$

The integrals appearing in this expression are readily performed, and we obtain the result [20]:

$$B_k^j(\infty) = \frac{(2j+k+1)!}{(2j+1)!k!} - 2^{k-1}\frac{(j+k+1)!}{j!k!}(1+\delta_{j,0}\delta_{k,0}) \quad . \tag{3.98}$$

The matrix elements are then [21]

$$J_{rr'}^l = \frac{\pi a^2}{\sigma}\left[\frac{r!r'!}{\Gamma(r+l+\frac{3}{2})\Gamma(r'+l+\frac{3}{2})}\right]^{\frac{1}{2}} 2^{-r-r'-l+\frac{3}{2}} l!$$

$$\times \sum_{j=0}^{r,r'}\sum_{k=0}^{l} \frac{4^j\Gamma(r+r'-2j+l-k-\frac{1}{2})}{\Gamma(-\frac{1}{2})(r-j)!(r'-j)!(l-k)!} B_k^j(\infty) \; . \tag{3.99}$$

The values of these matrix elements for the first few values of r and l are given in appendix A.

5. FURTHER PROPERTIES OF THE LINEARIZED COLLISION OPERATOR FOR HARD SPHERES

The only case for which all the eigenvalues and eigenfunctions of the linearized collision operator are known is that of Maxwell molecules; the eigenfunctions are the Burnett functions and the eigenvalues are given by (3.95). The spectrum of eigenvalues in this Maxwell case is entirely discrete, each eigenvalue having finite multiplicity, and, excepting the five-fold degenerate zero eigenvalue, the spectrum is entirely negative, with no lower bound. The question naturally arises to what extent is this spectrum typical of that for more general intermolecular force laws? Little is known with certainty about the answer, but we learn something from the case of hard spheres.

As we showed in section 3, the linearized collision operator for hard spheres may be written:

$$\boldsymbol{J}h = -m(c)h + \boldsymbol{K}h \;, \tag{3.100}$$

where the Hilbert operator $\boldsymbol{K}$ is the integral operator (3.33) with kernel given by (3.54) and the function $m(c)$ is given by (3.53). For c positive $m(c)$ is a positive definite monotonically increasing function of c with minimum value

$$m(0) = 2\pi^{\frac{1}{2}} a^2/\sigma \; . \tag{3.101}$$

For large c,

$$m(c) \sim \frac{\pi a^2}{\sigma} c \quad , \tag{3.102}$$

so $m(c)$ is unbounded.

For this case of hard spheres it has been shown that [22]

$$\int \dots \int \mathrm{d}c \dots \mathrm{d}c_5 \, \mathrm{e}^{-c^2 \dots -c_5^2} K(c, c_1) K(c_1, c_2) K(c_2, c_3) K(c_3, c_4)$$

$$\times K(c_4, c_5) K(c_5, c) < \infty \quad , \tag{3.103}$$

which means that the third iterate of the Hilbert operator $\boldsymbol{K}$ is of Hilbert-Schmidt type [23]. This implies in turn that the Hilbert operator is a completely continuous operator [24]. Such an operator has many simple properties: it is bounded with a completely discrete spectrum, each non-zero eigenvalue having finite multiplicity, and with zero the only limit point of the spectrum. In addition it has been shown long ago that $\boldsymbol{K}$ is a positive operator, so the eigenvalues of $\boldsymbol{K}$ are all positive [25].

But the spectrum of $\boldsymbol{K}$ is by no means the spectrum of $\boldsymbol{J}$. In fact, $\boldsymbol{K}$ differs from $\boldsymbol{J}$ by the operation of multiplication by the function $m(c)$. The spectrum of this operation is just the set of values $m(c)$ assumes as c takes on all possible values, i.e. the continuum from $m(0)$ to infinity [26]. We can now use a famous theorem of Weyl and von Neumann which states that when any self-adjoint operator is perturbed by adding to it a completely continuous operator the continuous spectrum of the first operator is unchanged [27]. In our case this means that the linearized collision operator has a continuous spectrum from $-m(0)$ to $-\infty$.

The Weyl theorem tells us nothing about the existence or non-existence of discrete eigenvalues of $\boldsymbol{J}$, nor does it tell us what are the eigenfunctions associated with the continuous spectrum. With regard to the discrete eigenvalues, it is easy to show that there are at least some in the gap between zero and $-m(0)$ [28]. Recently Kuščer and Williams have used an ingenious argument to show there are infinitely many eigenvalues in the gap [29]. Aside from this not much is known.

What about the spectrum of the collision operator for other intermolecular potentials ? Here almost nothing is known. For potentials which do not strictly vanish beyond some maximum intermolecular separation the classical cross section diverges so strongly at small angles that $m(c)$ does not exist and therefore the separation (3.100) cannot be made. If one truncates the cross section at small angles the separation can be made and one can attempt to apply the Weyl theorem to make some statement about the continuous spectrum [30]. Since a natural cutoff in the cross section is provided by quantum mechanics, we would speculate that the spectrum for, say, the Lennard-Jones potential, including the effects of quantum mechanics, is rather similar to that for the hard sphere model. Clearly, however, much work needs to be done before we can say that we understand the mathematical properties of these operators.

NOTES

1) For a brief introduction to the Boltzmann equation see: G. E. Uhlenbeck and G. W. Ford, *Lectures in Statistical Mechanics* (American Mathematical Society, Providence, Rhode Island, 1963) esp. chapter IV.

2) Describing the motion by classical mechanics, we have

$$\sin\theta\, I(u,\theta) = p\left|\mathrm{d}p/\mathrm{d}\theta\right| \quad ,$$

where $p(u,\theta)$, the impact parameter, is given implicitly by the relation:

$$\theta = \pi - 2\int_0^{\eta_0} \mathrm{d}\eta \left[1 - \eta^2 - \frac{4}{mu^2}\varphi\!\left(\frac{p}{\eta}\right)\right]^{-\frac{1}{2}} \quad .$$

Here $\varphi(r)$ is the intermolecular force potential and m is the molecular mass, while η_0 is the smallest (positive) zero of the expression within the square bracket. For a derivation of these formulas, see, e.g., L. D. Landau and E. M. Lifshitz, *Mechanics* (Pergamon Press, London, 1960) chapter IV.

For elastic spheres of diameter a, for which

$$\varphi(r) = \begin{cases} 0 , & a < r , \\ \infty , & 0 < r < a , \end{cases}$$

we find

$$I(u,\theta) = \tfrac{1}{4}a^2 \quad .$$

Another case of interest is that of Maxwell molecules, which repel each other with a force proportional to the inverse fifth power of the distance, i.e.,

$$\varphi(r) = \varphi_4 \left(\frac{a}{r}\right)^4, \qquad \varphi_4 > 0 .$$

Here we can write

$$\sin\theta\, I(u,\theta) = a^2 \left(\frac{\varphi_4}{mu^2}\right)^{\frac{1}{2}} \frac{(\cos 2\psi)^{\frac{1}{2}}}{\sin 2\psi[\cos^2\psi K(\sin\psi) - \cos 2\psi E(\sin\psi)]},$$

where

$$\theta = \pi - 2\,(\cos 2\psi)^{\frac{1}{2}} K(\sin\psi) .$$

In these expressions

$$K(x) = \int_0^{\frac{1}{2}\pi} d\alpha\,(1 - x^2 \sin^2\alpha)^{-\frac{1}{2}} ,$$

and

$$E(x) = \int_0^{\frac{1}{2}\pi} d\alpha\,(1 - x^2 \sin^2\alpha)^{\frac{1}{2}}$$

are, respectively, the complete elliptic integrals of the first and second kind. These expressions were first obtained by J. C. Maxwell, *Scientific Papers* (ed. W. D. Niven, Cambridge, University Press, 1890) p. 42.

It is interesting to remark that in general for a repulsive power law potential of the form

$$\varphi(r) = \varphi_s \left(\frac{a}{r}\right)^s ,$$

the cross section is of the form

$$I(u,\theta) = a^2 \left(\frac{4\varphi_s}{mu^2}\right)^{\frac{2}{s}} f_s(\theta)$$

where $f_s(\theta)$ is a function of θ alone. In fact, for small angles θ, it can be shown in general that

$$f_s(\theta) \approx \frac{1}{s}\left[\frac{\Gamma\left(\frac{s+1}{2}\right)\Gamma(\frac{1}{2})}{\Gamma(\frac{s}{2})}\right]^{\frac{2}{s}} \theta^{-2-\frac{2}{s}} .$$

(see, e.g., Landau and Lifshitz, *op. cit.*)

The cross sections for more general potentials must be calculated numerically. See, e.g., Hirschfelder, Curtis and Bird, *Molecular Theory of Gases and Liquids* (John Wiley and Sons, Inc., New York, 1964) pp. 533-9.

3) For some discussion of the quasi-classical limit of the scattering cross sections see, e.g., L. D. Landau and E. M. Lifshitz, *Quantum Mechanics*, (Pergamon Press, London, 1958) pp. 414-7.

4) An exception is ^{3}He or ^{4}He at low temperatures. For a discussion of quantum effects see, e.g., the article by J. de Boer and R. B. Bird in Hirschfelder, Curtis and Bird, *op. cit.*, chapter 10.

5) Note especially that this is *not* the assumption made in the Chapman-Enskog development where one linearizes about a local Maxwell distribution, for which the density, temperature, and local flow velocity are function of $\boldsymbol{r}$ and t. See, e.g., Uhlenbeck and Ford, *op. cit.*, chapter VI.

6) Thus

$$(\chi, \boldsymbol{J}\varphi) = \pi^{-3} \int \mathrm{d}\boldsymbol{c} \int \mathrm{d}\boldsymbol{c}_1 \, e^{-c^2 - c_1^2} \int \mathrm{d}\Omega \, F(g,\theta) \, \chi[\varphi' + \varphi_1' - \varphi - \varphi_1]$$

$$= \pi^{-3} \int \mathrm{d}\boldsymbol{c} \int \mathrm{d}\boldsymbol{c}_1 \, e^{-c^2 - c_1^2} \int \mathrm{d}\Omega \, F(g,\theta) \chi_1[\varphi' + \varphi_1' - \varphi - \varphi_1] \, ,$$

by interchanging $\boldsymbol{c}$ and $\boldsymbol{c}_1$. Forming half the sum of these two expressions;

$$(\chi, \boldsymbol{J}\varphi) = \frac{1}{2\pi^3} \int \mathrm{d}\boldsymbol{c} \int \mathrm{d}\boldsymbol{c}_1 \, e^{-c^2 - c_1^2} \int \mathrm{d}\Omega \, F(g,\theta)[\chi + \chi_1][\varphi' + \varphi_1' - \varphi - \varphi$$

$$= \frac{1}{2\pi^3} \int \mathrm{d}\boldsymbol{c} \int \mathrm{d}\boldsymbol{c}_1 \, e^{-c^2 - c_1^2} \int \mathrm{d}\Omega \, F(g,\theta)[\chi' + \chi_1'][\varphi + \varphi_1 - \varphi' -$$

Here this second expression results from interchanging $\boldsymbol{c}$ with $\boldsymbol{c}'$ and $\boldsymbol{c}_1$ with $\boldsymbol{c}_1'$ and using the facts that $c^2 + c_1^2 = c'^2 + c_1'^2$ and that $\mathrm{d}\boldsymbol{c}\,\mathrm{d}\boldsymbol{c}_1 = \mathrm{d}\boldsymbol{c}'\,\mathrm{d}\boldsymbol{c}_1'$. If we form half the sum of these last two expressions we get the expression in the text.

7) The book of F. Riesz and B. Sz.-Nagy, *Functional Analysis* (Frederick Ungar Publishing Co., New York, 1955) is a standard reference work on the mathematics of linear operators. In this case we refer to section 122.

8) The spherical harmonics are defined for positive integer values of l and for $m = 0, \pm 1, \pm 2, \ldots, \pm l$. If α and β are, resp., the polar and azimuthal angles of the unit vector $\hat{\boldsymbol{c}}$, then

$$Y_{lm}(\hat{c}) = (-)^m \left[\frac{2l+1}{4\pi} \frac{(l-m)!}{(l+m)!}\right]^{\frac{1}{2}} P_l^m(\cos\alpha)\, e^{im\beta} \quad ,$$

where

$$P_l^m(x) = \frac{(1-x^2)^{\frac{1}{2}m}}{2^l l!} \frac{d^{l+m}}{dx^{l+m}} (x^2-1)^l$$

is the associated Legendre polynomial. The spherical harmonics are normalized to unity:

$$\int d\omega\, Y_{lm}^*(\hat{c})\, Y_{l'm'}(\hat{c}) = \delta_{ll'}\, \delta_{mm'} \quad ,$$

where the integral is over all directions of $\hat{c}$ with $d\omega = \sin\alpha\, d\alpha\, d\beta$ the element of solid angle. See, e.g., A. R. Edmonds, *Angular Momentum in Quantum Mechanics* (Princeton University Press, Princeton, 1957) esp. section 2.5.

9) See David Enskog, *Kinetische Theorie der Vorgänge in Mässig Verdünnten Gasen* (Dissertation, Uppsala, 1917) pp. 140-148. Enskog's expression was a generalization of an earlier form obtained by Hilbert for the special case of hard spheres. See D. Hilbert, *Grundzüge einer Allgemeinen Theorie der Linearen Integralgleichungen* (B. G. Teubner, Leipzig, 1924) pp. 272-276. The proof given in the text is a slight modification of that of L. Waldmann, *Handbuch der Physik,* Vol. XII (Springer-Verlag, Berlin, 1958) pp. 366-367.

10) See, e.g., W. Magnus and F. Oberhettinger, *Formulas and Theorems for the Special Functions of Mathematical Physics* (Chelsea Publishing Co., New York, 1949) p. 19.

11) In general, if $f(x_0) = 0$, $\delta[f(x)] = |f'(x_0)|^{-1} \delta(x - x_0)$.

12) The Burnett functions were first introduced by D. Burnett, Proc. Lond. Math. Soc. 39 (1935) 385. They are used extensively by Chapman and Cowling, *The Mathematical Theory of Non-Uniform Gases* 2nd edition (Cambridge Univerity Press, Cambridge, 1958).

13) Magnus and Oberhettinger, *op. cit.*, p. 84.

14) Edmonds, *op. cit.*, p. 63.

15) To obtain this identity we begin with the integral representation of Laplace and Mehler (see Magnus and Oberhettinger, *op. cit.*, p. 52):

$$P_l(\cos\beta) = \frac{1}{2\pi} \int_0^{2\pi} d\alpha\, (\cos\beta + i \sin\beta \cos\alpha)^l \ .$$

Hence,

$$(cc_1)^l P_l(\hat{c}\cdot\hat{c}_1) = \frac{1}{2\pi}\int_0^{2\pi} d\alpha\,(c\cdot c_1 + i\,|c\times c_1|\cos\alpha)^l ,$$

and

$$\sum_{l=0}^{\infty}(cc_1)^l P_l(\hat{c}\cdot\hat{c}_1)\frac{t^l}{l!} = \frac{1}{2\pi}\int_0^{2\pi} d\alpha\, e^{tc\cdot c_1 + it|c\times c_1|\cos\alpha} .$$

The identity (3.67) follows from the fact that if $\hat{l}$ is a unit vector lying in the plane perpendicular to $c - c_1$, then

$$(\hat{l}\cdot(c\times c_1)) = |c\times c_1|\cos\alpha ,$$

where α is the angle between $c\times c_1$ and $\hat{l}$.

16) Magnus and Oberhettinger, *op. cit.*, p. 26.

17) Chapman and Cowling, *op. cit.*, p. 157. The coefficients have been chosen so these Ω-integrals are identical with those of Chapman and Cowling.

The relation between the matrix elements we introduce and the bracket expressions of Chapman and Cowling is

$$\left[S_{\frac{3}{2}}^{(r)}(\mathcal{C}^2)\mathbf{\mathcal{C}}\,,\ S_{\frac{3}{2}}^{(r')}(\mathcal{C}^2)\mathbf{\mathcal{C}}\right] = -\tfrac{15}{4}\sigma\left(\frac{2kT}{m}\right)^{\frac{1}{2}} J^1_{rr'} ,$$

$$\left[S_{\frac{5}{2}}^{(r)}(\mathcal{C}^2)\mathbf{\mathcal{C}}^{\circ}\mathbf{\mathcal{C}}\,,\ S_{\frac{5}{2}}^{(r')}(\mathcal{C}^2)\mathbf{\mathcal{C}}^{\circ}\mathbf{\mathcal{C}}\right] = -\tfrac{5}{2}\sigma\left(\frac{2kT}{m}\right)^{\frac{1}{2}} J^2_{rr'} .$$

18) A table of the first 559 eigenvalues for the case of Maxwell molecules is given by Z. Alterman, K. Frankowski and C. L. Pekeris, Astrophysical J. Suppl. Series 7 (1962) 291. In their expressions one must put $F(\theta) = 2^{-\frac{1}{2}} f_4(\theta)$ and

$$\lambda_{rl} = \frac{2a^2}{\sigma}\left(\frac{\varphi_4}{kT}\right)^{\frac{1}{2}}(\lambda_{rl})_{\mathrm{AFP}} ,$$

where $(\lambda_{rl})_{\mathrm{AFP}}$ are the tabulated numerical values of Alterman, Frankowski and Pekeris. The definition of $F(\theta)$ and (λ_{rl})AFP are identical with those of C. S. Wang Chang and G. E. Uhlenbeck, *On the Progagation of Sound in Monoatomic Gases* (this *Study*).

The fact that the Burnett functions are eigenfunctions of the collision operator was first recognized by Wang Chang and Uhlenbeck, *op. cit.*, and, independently, by H. M. Mott-Smith, *A New Approach to the Kinetic Theory of Gases* (Lincoln Laboratory Group Report V-2, 1954). See also L. Waldmann, *op. cit.*, section 38.

19) See Wang Chang and Uhlenbeck, *op. cit.*, appendix III.

20) In the first term we replace θ by $x = \cos(\frac{1}{2}\theta)$ as variable of integration, and use the integration formula:

$$\int_0^1 \mathrm{d}x\, x^{2j+k+1} P_k(x) = 2^{-k-1} \frac{(2j+k+1)!\, j!}{(2k+1)!(j+k+1)!} \quad .$$

See Magnus and Oberhettinger *op.cit.*, p. 52. The second term is similar and the last two terms are elementary.

21) This formula for the matrix elements for elastic spheres was first given by Mott-Smith, *op. cit.*, note 18. This result is quoted by C. L. Pekeris, Z. Alterman, L. Finkelstein and K. Frankowski, Phys. Fluids 5 (1962) 1608, where the matrix elements are expressed in terms of a symbol $[rlr'l]$. The relation with our notation is

$$J^l_{rr'} = -\frac{\Gamma(\frac{3}{2})}{\sigma}(2l+1)\left[\frac{r!r'!}{\Gamma(r+l+\frac{3}{2})\Gamma(r'+l+\frac{3}{2})}\right]^{\frac{1}{2}} [rlr'l] \; .$$

Extensive tables of these matrix elements, still for the case of elastic spheres, are given by L. Sirovich and J. K. Thurber, *Advances in Applied Mechanics*, Suppl. 3, Vol. 1 (Rarefied Gas Dynamics) 1965, ed. J. H. Leeuw, p. 21. They introduce the symbol $B_{rl;r'l'}$ which is related to our $J^l_{rr'}$ by

$$J^l_{rr'} = \frac{4\pi}{\sigma} B_{rl;r'l} \; .$$

22) See J. R. Dorfman, Proc. Nat. Acad. Sci. 50 (1963) 804. Also H. Grad, in: *Proceedings of the Third International Symposium on Rarefield Gas Dynamics* ed. J.A. Laurmann (Academic Press, New York, 1963) pp. 26-59.

23) See Riesz and Sz.-Nagy, *op. cit.*, section 133.

24) See Riesz and Sz.-Nagy, *op. cit.*, section 77.

25) See E. Hecke, Math. Z. 12 (1922) 274.

26) Multiplication by a function, considered as an operation, has "eigenfunctions" which are Dirac delta functions and eigenvalues which are the values of the function. Thus

$$m(c)\,\delta(c - c_0) = m(c_0)\delta(c - c_0) \; .$$

Here $m(c_0)$ is the eigenvalue. Of course the delta function is not strictly an eigenfunction since it is not even a function, much less a square integrable function.

27) See Riesz and Sz.-Nagy, *op. cit.*, section 134.

28) The proof that there is at least one eigenvalue in the gap consists in noting that since for hard spheres

$$J^1_{11} = -\frac{a^2}{\sigma}\frac{16(2\pi)^{\frac{1}{2}}}{15}$$

we have

$$-m(0) < J^1_{11} < 0 \ ,$$

and, therefore, by the Rayleigh-Ritz principle there must be at least one eigenfunction with eigenvalue greater than J^1_{11}. See G. W. Ford and M. Schreiber, Proc. Nat. Acad. Sci. 60 (1968) 802.

29) I. Kuščer and M. M. R. Williams, Phys. Fluids 10 (1967) 1922. See also C. C. Yan, Phys. Fluids 12 (1969) 2306.

30) See, e.g., H. Grad, *op. cit*, note 22.

Chapter IV

THE DISPERSION OF SOUND FROM THE LINEARIZED BOLTZMANN EQUATION

1. INTRODUCTION

In chapter II we discussed what we might call the orthodox way of calculating the dispersion law for sound propagation. This involves first deriving hydrodynamic equations from the Boltzmann equation, using the Chapman-Enskog development, and then seeking normal mode solutions of the linearized hydrodynamic equations. In 1952 Mott-Smith and, independently, Wang Chang and Uhlenbeck pointed out that the intermediate step of deriving the hydrodynamic equations is unnecessary, one can seek normal mode solutions of the linearized Boltzmann equation directly [1]. Since the report of Chang and Uhlenbeck is included in this present volume, we give only a very brief review of their method in the next section.

The method of Mott-Smith and Chang and Uhlenbeck leads to an expansion of the dispersion law in powers of the ratio of mean free path to the wave length. The main aim of this chapter will be to obtain this same expansion directly by using standard methods of perturbation theory.

2. THE METHOD OF MOTT-SMITH, CHANG and UHLENBECK

We begin with the linearized Boltzmann equation (3.14), which we write in the form:

$$\left(\frac{m}{2\,kT}\right)^{\frac{1}{2}} \frac{\partial h}{\partial t} + \boldsymbol{c}\cdot\frac{\partial h}{\partial \boldsymbol{r}} = n\sigma \boldsymbol{J} h \,. \tag{4.1}$$

Here $\boldsymbol{c}$ is the dimensionless velocity variable introduced in (3.16) and we have in mind that the linearized collision operator $\boldsymbol{J}$ is in the dimensionless form (3.17). We seek normal mode solutions of this equation of the form:

$$h(\boldsymbol{r}, \boldsymbol{c}, t) = h_{\omega,\boldsymbol{k}}(\boldsymbol{c})\, e^{i(\boldsymbol{k}\cdot\boldsymbol{r}-\omega t)} . \tag{4.2}$$

Putting this in the equation (4.1) gives:

$$\left[-i\left(\frac{m}{2\,kT}\right)^{\frac{1}{2}} \omega + i\boldsymbol{k}\cdot\boldsymbol{c}\right] h_{\omega,\boldsymbol{k}} = n\sigma \boldsymbol{J} h_{\omega,\boldsymbol{k}} , \tag{4.3}$$

which is the basic equation we discuss. We can simplify this equation a bit if we recognize that sound, the normal mode we want to study, is a longitudinal mode of oscillation. Such a mode is axially symmetric about the direction of propagation. If, therefore, we choose the z-axis to be along $\boldsymbol{k}$, we can write (4.3) in the form:

$$(E - \epsilon\, c_z) h_{\omega,k} = \boldsymbol{J} h_{\omega,k} , \tag{4.4}$$

where

$$E = -i\left(\frac{m}{2\,kT}\right)^{\frac{1}{2}} \frac{\omega}{n\sigma} , \quad \epsilon = -i\,\frac{k}{n\sigma} . \tag{4.5}$$

In (4.4) we have in mind that we seek only axially symmetric solutions $h_{\omega,k}(c, c_z)$, which depend only upon c, the magnitude of $\boldsymbol{c}$, and c_z, the component of $\boldsymbol{c}$ along $\boldsymbol{k}$.

Mott-Smith and Chang and Uhlenbeck begin the discussion of this equation by expanding $h_{\omega,k}$ in terms of the axially symmetric Burnett functions. These are

$$\psi_{rl}(c, c_z) = \psi_{rl0}(\boldsymbol{c}) , \tag{4.6}$$

where $\psi_{rlm}(\boldsymbol{c})$ is the normalized Burnett function defined in (3.58). For convenience the first few ψ_{rl} are given in table 2. In (4.4) we put, therefore,

$$h_{\omega,k} = \sum_{r=0}^{\infty} \sum_{l=0}^{\infty} a_{rl}\, \psi_{rl} , \tag{4.7}$$

and form the scalar product of both sides of the equation with ψ_{rl}. The result is an infinite set of linear equations for the determination of the expansion coefficients a_{rl}, the general equation being:

$$E\, a_{rl} - \epsilon \sum_{r'=0}^{\infty} \sum_{l'=0}^{\infty} (\psi_{rl}, c_z \psi_{r'l'}) a_{r'l'}$$

$$- \sum_{r'=0}^{\infty} J^{l}_{rr'}\, a_{r'l} = 0 . \tag{4.8}$$

Table 2
The first eleven axially symmetric Burnett functions

$$\psi_{00} = 1$$
$$\psi_{01} = (2)^{\frac{1}{2}} c_z$$
$$\psi_{10} = (\tfrac{2}{3})^{\frac{1}{2}} (\tfrac{3}{2} - c^2)$$
$$\psi_{11} = (\tfrac{4}{5})^{\frac{1}{2}} (\tfrac{5}{2} - c^2) c_z$$
$$\psi_{02} = (\tfrac{1}{3})^{\frac{1}{2}} (3 c_z^2 - c^2)$$
$$\psi_{03} = (\tfrac{2}{15})^{\frac{1}{2}} (5 c_z^2 - 3 c^2) c_z$$
$$\psi_{20} = (\tfrac{2}{15})^{\frac{1}{2}} (\tfrac{15}{4} - 5 c^2 + c^4)$$
$$\psi_{12} = (\tfrac{2}{21})^{\frac{1}{2}} (\tfrac{7}{2} - c^2)(3 c_z^2 - c^2)$$
$$\psi_{04} = (\tfrac{1}{420})^{\frac{1}{2}} (35 c_z^4 - 30 c_z^2 c^2 + 3 c^4)$$
$$\psi_{21} = (\tfrac{4}{35})^{\frac{1}{2}} (\tfrac{35}{4} - 7 c^2 + c^4) c_z$$
$$\psi_{13} = (\tfrac{4}{135})^{\frac{1}{2}} (\tfrac{9}{2} - c^2)(5 c_z^2 - 3 c^2) c_z$$

Here we have used the expression (3.59) for the matrix elements of $\boldsymbol{J}$ in terms of the symbol $J^l_{rr'}$. The matrix elements of c_z which appear are relatively simple. We can evaluate them with the help of the recursion relation [2]:

$$c_z \psi_{rl} = (l+1)\left[\frac{r+l+\frac{3}{2}}{(2l+3)(2l+1)}\right]^{\frac{1}{2}} \psi_{r,l+1} - (l+1)\left[\frac{r}{(2l+3)(2l+1)}\right]^{\frac{1}{2}} \psi_{r-1,l+1}$$

$$+\, l\left[\frac{r+l+\frac{1}{2}}{(2l+1)(2l-1)}\right]^{\frac{1}{2}} \psi_{r,l-1} - l\left[\frac{r+1}{(2l+1)(2l-1)}\right]^{\frac{1}{2}} \psi_{r+1,l-1}\,. \tag{4.9}$$

Therefore, the matrix elements of c_z are

(r,l) \ (r',l')	(0,0)	(0,1)	(1,0)	(0,2)	(1,1)	(0,3)	(2,0)	(1,2)	(0,4)	(2,1)	(1,3)
(0,0)	E	$-(\frac{1}{2})^{\frac{1}{2}}\epsilon$	0	0	0	0	0	0	0	0	0
(0,1)	$-(\frac{1}{2})^{\frac{1}{2}}\epsilon$	E	$(\frac{1}{3})^{\frac{1}{2}}\epsilon$	$-(\frac{2}{3})^{\frac{1}{2}}\epsilon$	0	0	0	0	0	0	0
(1,0)	0	$(\frac{1}{3})^{\frac{1}{2}}\epsilon$	E	0	$-(\frac{5}{6})^{\frac{1}{2}}\epsilon$	0	0	0	0	0	0
(0,2)	0	$-(\frac{2}{3})^{\frac{1}{2}}\epsilon$	0	$E-J^2_{00}$	$(\frac{4}{15})^{\frac{1}{2}}\epsilon$	$-(\frac{9}{10})^{\frac{1}{2}}\epsilon$	0	$-J^2_{01}$	0	0	0
(1,1)	0	0	$-(\frac{5}{6})^{\frac{1}{2}}\epsilon$	$(\frac{4}{15})^{\frac{1}{2}}\epsilon$	$E-J^1_{11}$	0	$(\frac{2}{3})^{\frac{1}{2}}\epsilon$	$-(\frac{14}{15})^{\frac{1}{2}}\epsilon$	0	$-J^1_{12}$	0
(0,3)	0	0	0	$-(\frac{9}{10})^{\frac{1}{2}}\epsilon$	0	$E-J^3_{00}$	0	$(\frac{9}{35})^{\frac{1}{2}}\epsilon$	$-(\frac{8}{7})^{\frac{1}{2}}\epsilon$	0	$-J^3_{01}$
(2,0)	0	0	0	0	$(\frac{2}{3})^{\frac{1}{2}}\epsilon$	0	$E-J^0_{22}$	0	0	$-(\frac{7}{6})^{\frac{1}{2}}\epsilon$	0
(1,2)	0	0	0	$-J^2_{10}$	$-(\frac{14}{15})^{\frac{1}{2}}\epsilon$	$(\frac{9}{35})^{\frac{1}{2}}\epsilon$	0	$E-J^2_{11}$	0	$(\frac{8}{15})^{\frac{1}{2}}\epsilon$	$-(\frac{81}{70})^{\frac{1}{2}}\epsilon$
(0,4)	0	0	0	0	0	$-(\frac{8}{7})^{\frac{1}{2}}\epsilon$	0	0	$E-J^4_{00}$	0	$(\frac{16}{63})^{\frac{1}{2}}\epsilon$
(2.1)	0	0	0	0	$-J^1_{21}$	0	$-(\frac{7}{6})^{\frac{1}{2}}\epsilon$	$(\frac{8}{15})^{\frac{1}{2}}\epsilon$	0	$E-J^1_{22}$	0
(1,3)	0	0	0	0	0	$-J^3_{10}$	0	$-(\frac{81}{70})^{\frac{1}{2}}\epsilon$	$(\frac{16}{63})^{\frac{1}{2}}\epsilon$	0	$E-J^3_{11}$

$$(\psi_{rl}, c_z \psi_{r'l'}) = (l+1) \left[\frac{r+l+\frac{3}{2}}{(2l+3)(2l+1)}\right]^{\frac{1}{2}} \delta_{r,r'} \delta_{l+1,l'}$$

$$- (l+1) \left[\frac{r}{(2l+3)(2l+1)}\right]^{\frac{1}{2}} \delta_{r-1,r'} \delta_{l+1,l'}$$

$$+ l \left[\frac{r+l+\frac{1}{2}}{(2l+1)(2l-1)}\right]^{\frac{1}{2}} \delta_{r,r'} \delta_{l-1,l'}$$

$$- l \left[\frac{r+1}{(2l+1)(2l-1)}\right]^{\frac{1}{2}} \delta_{r+1,r'} \delta_{l-1,l'} \; . \tag{4.10}$$

In order that there be a non-trivial solution of the equations (4.8) it is necessary that the associated infinite determinant vanish [3]. The upper left hand corner of this determinant, which we call the Chang determinant, is shown on the opposite page. Setting this determinant equal to zero gives us the dispersion relation we desire. Chang and Uhlenbeck approximate this infinite determinant by successive truncations, the simplest scheme being that indicated by the heavy lines. Thus, in the 0'th approximation the 3 by 3 determinant in the upper left hand corner is set equal to zero to give

$$E(E^2 - \tfrac{5}{6}\epsilon^2) = 0 \; . \tag{4.11}$$

The two solutions:

$$E = \pm \left(\tfrac{5}{6}\right)^{\frac{1}{2}} \epsilon \; , \tag{4.12}$$

correspond to the sound modes, and the solution:

$$E = 0 \; , \tag{4.13}$$

corresponds to the heat conduction mode. This 0'th approximation clearly corresponds to the Euler approximation to hydrodynamics.

The next approximation, the first approximation, is obtained by setting the 5×5 determinant in the upper left hand corner equal to zero to get

$$E(E^2 - \tfrac{5}{6}\epsilon^2)[(E - J^1_{11})(E - J^2_{00}) - \tfrac{4}{15}\epsilon^2]$$

$$- \tfrac{5}{6}\epsilon^2(E^2 - \tfrac{1}{2}\epsilon^2)(E - J^2_{00}) - \tfrac{2}{3}\epsilon^2 E^2(E - J^1_{11})$$

$$+ \tfrac{1}{9} E\epsilon^4 = 0 \ . \tag{4.14}$$

This is a polynomial of fifth degree in E, so there are five zeros, three corresponding to the zeros obtained in the 0'th approximation and two new ones. For small ϵ we can expand these zeros in powers of ϵ to get:

$$E = \pm \left(\tfrac{5}{6}\right)^{\frac{1}{2}} \epsilon - \left[\frac{1}{6J^1_{11}} + \frac{1}{3J^2_{00}}\right]\epsilon^2 + \dots \ ,$$

$$E = -\frac{1}{2J^1_{11}}\epsilon^2 + \dots \ ,$$

$$E = J^1_{11} + \dots \ ,$$

$$E = J^2_{00} + \dots \ . \tag{4.15}$$

The first pair of zeros can be identified as those corresponding to the sound mode as obtained from the Navier-Stokes equations of hydrodynamics in which the viscosity and heat conduction are given in the first Enskog approximation [4]:

$$[\eta]_1 = -\frac{m}{2\sigma J^2_{00}}\left(\frac{2kT}{m}\right)^{\frac{1}{2}} , \quad [\kappa]_1 = -\frac{5k}{4\sigma J^1_{11}}\left(\frac{2kT}{m}\right)^{\frac{1}{2}} \ . \tag{4.16}$$

Similarly the third zero is that corresponding to the heat conduction mode in the same approximation. The last two zeros correspond to highly damped modes and, in the light of what we shall learn about the normal modes in chapters V and VI, should be regarded as spurious.

If we go on to the second approximation where we set the 8×8 determinant in the upper left hand corner equal to zero, we get a polynomial of eighth degree in E. There are eight zeros, five of them spurious and three which when expanded in powers of ϵ correspond to the sound and heat conduction modes in the Burnett approximation to hydrodynamics.

As one goes on in this way, one obtains an expansion of E in powers of ϵ for the sound and heat conduction modes, the successive orders corresponding to higher order approximations to the hydrodynamical equations [5]. Although this method is more direct than that of first deriving the successive hydrodynamical equations, linearizing, and then seeking the normal modes, it still becomes soon very complicated. Since the result is an expansion of E in powers of ϵ, it would seem to be still simpler to obtain this expansion directly from the linearized Boltzmann equation by perturbation methods. We show how this can be done in the next section.

3. THE PERTURBATION METHOD

We begin again with equation (4.4) which we write in the form

$$J h_{\omega,k} + \epsilon\, c_z\, h_{\omega,k} = E h_{\omega,k} , \tag{4.17}$$

where, to repeat,

$$\epsilon = -i \frac{k}{n\sigma} , \qquad E = -i \left(\frac{m}{2kT}\right)^{\frac{1}{2}} \frac{\omega}{n\sigma} . \tag{4.18}$$

Since we want to obtain the dispersion law in the form of an expansion of ω in powers of k, we write

$$h_{\omega,k} = h^{(0)}_{\omega,k} + \epsilon h^{(1)}_{\omega,k} + \epsilon^2 h^{(2)}_{\omega,k} + \dots ,$$

$$E = \epsilon E^{(1)} + \epsilon^2 E^{(2)} + \dots . \tag{4.19}$$

Here we put $E^{(0)}$ equal to zero since for the sound and heat conduction modes $\omega = 0$ when $k = 0$, i.e. the frequency goes to zero at infinite wave lengths.

Inserting the expansions (4.19) in (4.17) and equating the coefficients of equal powers of ϵ on either side gives the following sequence of equations:

$$J h^{(0)}_{\omega,k} = 0 , \tag{4.20a}$$

$$\boldsymbol{J}h^{(1)}_{\omega,k} = (E^{(1)} - c_z)\,h^{(0)}_{\omega,k} \ , \tag{4.20b}$$

$$\boldsymbol{J}h^{(2)}_{\omega,k} = (E^{(1)} - c_z)\,h^{(1)}_{\omega,k} + E^{(2)}\,h^{(0)}_{\omega,k} \ , \tag{4.20c}$$

$$\boldsymbol{J}h^{(3)}_{\omega,k} = (E^{(1)} - c_z)\,h^{(2)}_{\omega,k} + E^{(2)}\,h^{(1)}_{\omega,k} + E^{(3)}\,h^{(0)}_{\omega,k} \ , \tag{4.20d}$$

and so on.

In 0'th order it follows from (4.20a) and the axial symmetry of the problem that $h^{(0)}_{\omega,k}$ must be a linear combination of the conserved quantities 1, c_z and c^2. The correct linear combination will be determined in first order. Here we put

$$h^{(0)}_{\omega,k} = A^{(0)}_1\,\varphi_1 + A^{(0)}_2\,\varphi_2 + A^{(0)}_3\,\varphi_3 \ , \tag{4.21}$$

where

$$\varphi_1 = (\tfrac{2}{15})^{\frac{1}{2}}\,c^2 + c_z \ , \tag{4.22a}$$

$$\varphi_2 = (\tfrac{2}{5})^{\frac{1}{2}}\,(\tfrac{5}{2} - c^2) \ , \tag{4.22b}$$

$$\varphi_3 = (\tfrac{2}{15})^{\frac{1}{2}}\,c^2 - c_z \ , \tag{4.22c}$$

and the $A^{(0)}$ coefficients are still to be determined.

The functions φ_j, $j = 1, 2, 3$, have been constructed so that they are orthonormal combinations of the conserved quantities which (with an eye toward the next order) diagonalize the perturbation c_z. In fact, using table 2 of the axially symmetric Burnett functions, we can verify directly that

$$c_z\,\varphi_1 = (\tfrac{5}{6})^{\frac{1}{2}}\,\varphi_1 - (\tfrac{1}{6})^{\frac{1}{2}}\,\psi_{11} + (\tfrac{1}{3})^{\frac{1}{2}}\,\psi_{02} \ ,$$

$$c_z\,\varphi_2 = (\tfrac{1}{2})^{\frac{1}{2}}\,\psi_{11} \ ,$$

$$c_z\,\varphi_3 = -\,(\tfrac{5}{6})^{\frac{1}{2}}\,\varphi_3 - (\tfrac{1}{6})^{\frac{1}{2}}\,\psi_{11} - (\tfrac{1}{3})^{\frac{1}{2}}\,\psi_{02} \ . \tag{4.23}$$

Note that $\varphi_1, \varphi_2, \varphi_3$, being linear combinations of $\psi_{00}, \psi_{01}, \psi_{10}$, are orthogonal to all the other Burnett functions.

In the first order we put the zeroth order solution (4.21) into (4.20b) to get

$$\boldsymbol{J} h^{(1)}_{\omega,k} = (E^{(1)} - c_z)(A_1^{(0)}\varphi_1 + A_2^{(0)}\varphi_2 + A_3^{(0)}\varphi_3) , \tag{4.24}$$

which is an inhomogeneous linear equation for the determination of $h^{(1)}_{\omega,k}$. Such an equation can be solved only if the inhomogeneous term is orthogonal to all the solutions of the homogeneous equation. Using (4.23), one finds three possible ways to satisfy this solvability criterion:

$$E^{(1)} = (\tfrac{5}{6})^{\frac{1}{2}} , \quad h^{(0)}_{\omega,k} = \varphi_1 , \tag{4.25a}$$

$$E^{(1)} = 0 \quad , \quad h^{(0)}_{\omega,k} = \varphi_2 , \tag{4.25b}$$

$$E^{(1)} = -(\tfrac{5}{6})^{\frac{1}{2}} , \quad h^{(0)}_{\omega,k} = \varphi_3 . \tag{4.25c}$$

The first and third possibilities correspond, resp., to sound modes propagating in the positive and negative z-directions (with the Laplace velocity $v_0 = (5kT/3m)^{\frac{1}{2}}$), while the second corresponds to the heat conduction mode. Since we are interested in the propagation of sound, we select the first possibility (4.25a). With this choice (4.24) becomes

$$\boldsymbol{J} h^{(1)}_{\omega,k} = \frac{1}{6^{\frac{1}{2}}}\psi_{11} - \frac{1}{3^{\frac{1}{2}}}\psi_{02} . \tag{4.26}$$

Note next that since $\boldsymbol{J}$ is a scalar operator, we can express $h^{(1)}_{\omega,k}$ in the form:

$$h^{(1)}_{\omega,k} = \sum_{r=1}^{\infty} a^{(1)}_{r1}\psi_{r1} + \sum_{r=0}^{\infty} a^{(1)}_{r2}\psi_{r2} + A_2^{(1)}\varphi_2 + A_3^{(1)}\varphi_3 . \tag{4.27}$$

Here we have imposed the requirement that $h^{(1)}_{\omega,k}$ be orthogonal to the zeroth order solution $h^{(0)}_{\omega,k}$, which accounts for the absence of a term proportional to φ_1. Indeed, we shall impose a similar requirement in all higher orders; it has no effect on the values of the coefficients $E^{(n)}$. We should note that the coefficients $A_2^{(1)}$ and $A_3^{(1)}$ cannot be determined from (4.26) since $\boldsymbol{J}$ annihilates φ_2 and φ_3. As we shall see, these coefficients will be determined in the next order.

To determine the coefficients $a_{r1}^{(1)}$ and $a_{r2}^{(1)}$, we insert the expression (4.27) in (4.26) and form the scalar products of both sides of the equation with respect to ψ_{r1} and ψ_{r2}. The result is two infinite sets of coupled linear equations:

$$\sum_{r'=1}^{\infty} J_{rr'}^{1} a_{r'1}^{(1)} = \left(\frac{1}{6}\right)^{\frac{1}{2}} \delta_{r,1} , \tag{4.28a}$$

$$\sum_{r'=0}^{\infty} J_{rr'}^{2} a_{r'2}^{(1)} = -\left(\frac{1}{3}\right)^{\frac{1}{2}} \delta_{r,0} . \tag{4.28b}$$

Here $J_{rr'}^{l}$ is the matrix element of $\boldsymbol{J}$ with respect to the Burnett functions defined in (3.59). (Remember $\psi_{rl} = \psi_{rl0}$.) We shall discuss the solution of these equations in the next two sections. For now we assume we have the solution and proceed to the next order.

Next consider the second order equation (4.20c). In the right hand side we put $h_{\omega,k}^{(0)}$ given by (4.25a) and $h_{\omega,k}^{(1)}$ given by (4.27). Again, this equation has a solution if and only if the right hand side is orthogonal to the solutions of the homogeneous equation, i.e., φ_1, φ_2 and φ_3. Forming the scalar product of the right hand side with each of these three functions, using (4.23), we get

$$E^{(2)} = -\left(\frac{1}{6}\right)^{\frac{1}{2}} a_{11}^{(1)} + \left(\frac{1}{3}\right)^{\frac{1}{2}} a_{02}^{(1)} ,$$

$$A_2^{(1)} = \left(\tfrac{3}{5}\right)^{\frac{1}{2}} a_{11}^{(1)} ,$$

$$A_3^{(1)} = -\left(\frac{1}{20}\right)^{\frac{1}{2}} a_{11}^{(1)} - \left(\frac{1}{10}\right)^{\frac{1}{2}} a_{02}^{(1)} . \tag{4.29}$$

Thus, the solvability criterion for the second order equation fixes the undertermined coefficients $A_2^{(1)}$ and $A_3^{(1)}$ in the first order solution and also determines $E^{(2)}$. With these results, and using (4.9), we can write the second order equation (4.20c) in the explicit form:

$$\boldsymbol{J} h^{(2)}_{\omega,k} = -(\tfrac{1}{3})^{\frac{1}{2}} \sum_{r=2}^{\infty} \left[(r+\tfrac{3}{2})^{\frac{1}{2}} a^{(1)}_{r1} - r^{\frac{1}{2}} a^{(1)}_{r-1,1} \right] \psi_{r0}$$

$$+ \sum_{r=1}^{\infty} \left\{ (\tfrac{5}{6})^{\frac{1}{2}} a^{(1)}_{r1} - (\tfrac{4}{15})^{\frac{1}{2}} \left[(r+\tfrac{5}{2})^{\frac{1}{2}} a^{(1)}_{r2} - r^{\frac{1}{2}} a^{(1)}_{r-1,2} \right] \right.$$

$$\left. - \left[(\tfrac{49}{120})^{\frac{1}{2}} a^{(1)}_{11} + (\tfrac{1}{60})^{\frac{1}{2}} a^{(1)}_{02} \right] \delta_{r,1} \right\} \psi_{r1}$$

$$+ \sum_{r=0}^{\infty} \left\{ (\tfrac{5}{6})^{\frac{1}{2}} a^{(1)}_{r2} - (\tfrac{4}{15})^{\frac{1}{2}} \left[(r+\tfrac{5}{2})^{\frac{1}{2}} a^{(1)}_{r1} - (r+1)^{\frac{1}{2}} a^{(1)}_{r+1,1} \right] \right.$$

$$\left. - \left[(\tfrac{1}{60})^{\frac{1}{2}} a^{(1)}_{11} + (\tfrac{1}{30})^{\frac{1}{2}} a^{(1)}_{02} \right] \delta_{r,0} \right\} \psi_{r2}$$

$$+ (\tfrac{9}{35})^{\frac{1}{2}} \sum_{r=0}^{\infty} \left[-(r+\tfrac{7}{2})^{\frac{1}{2}} a^{(1)}_{r2} + (r+1)^{\frac{1}{2}} a^{(1)}_{r+1,2} \right] \psi_{r3} \,. \qquad (4.30)$$

Again, since $\boldsymbol{J}$ is a scalar operator, we can express $h^{(2)}_{\omega,k}$ in the form:

$$h^{(2)}_{\omega,k} = \sum_{r=2}^{\infty} a^{(2)}_{r0} \psi_{r0} + \sum_{r=1}^{\infty} a^{(2)}_{r1} \psi_{r1}$$

$$+ \sum_{r=0}^{\infty} a^{(2)}_{r2} \psi_{r2} + \sum_{r=0}^{\infty} a^{(2)}_{r3} \psi_{r3}$$

$$+ A^{(2)}_{2} \varphi_{2} + A^{(2)}_{3} \varphi_{3} \,. \qquad (4.31)$$

Here, again, we have required that $h^{(2)}_{\omega,k}$ be orthogonal to φ_1, and the coefficients $A^{(2)}_2$ and $A^{(2)}_3$ must be determined from the third order equation.

The coefficients $a^{(2)}_{rl}$ occurring in (4.31) are determined by inserting (4.31) in (4.30) and forming the scalar product of both sides with ψ_{rl}. The result is four infinite sets of coupled linear equations:

$$\sum_{r'=2}^{\infty} J^{0}_{rr'} a^{(2)}_{r'0} = -(\tfrac{1}{3})^{\frac{1}{2}}\left[(r+\tfrac{3}{2})^{\frac{1}{2}} a^{(1)}_{r1} - r^{\frac{1}{2}} a^{(1)}_{r-1,1}\right], \tag{4.32a}$$

$$\sum_{r'=1}^{\infty} J^{1}_{rr'} a^{(2)}_{r'1} = (\tfrac{5}{6})^{\frac{1}{2}} a^{(1)}_{r1} - (\tfrac{4}{15})^{\frac{1}{2}}\left[(r+\tfrac{5}{2})^{\frac{1}{2}} a^{(1)}_{r2} - r^{\frac{1}{2}} a^{(1)}_{r-1,2}\right]$$

$$- \left[(\tfrac{49}{120})^{\frac{1}{2}} a^{(1)}_{11} + (\tfrac{1}{60})^{\frac{1}{2}} a^{(1)}_{02}\right]\delta_{r,1}, \tag{4.32b}$$

$$\sum_{r'=0}^{\infty} J^{2}_{rr'} a^{(2)}_{r'2} = (\tfrac{5}{6})^{\frac{1}{2}} a^{(1)}_{r2} - (\tfrac{4}{15})^{\frac{1}{2}}\left[(r+\tfrac{5}{2})^{\frac{1}{2}} a^{(1)}_{r1} - (r+1)^{\frac{1}{2}} a^{(1)}_{r+1,1}\right]$$

$$- \left[(\tfrac{1}{60})^{\frac{1}{2}} a^{(1)}_{11} + (\tfrac{1}{30})^{\frac{1}{2}} a^{(1)}_{02}\right]\delta_{r,0}, \tag{4.32c}$$

$$\sum_{r'=0}^{\infty} J^{3}_{rr'} a^{(2)}_{r'3} = (\tfrac{9}{35})^{\frac{1}{2}}\left[-(r+\tfrac{7}{2})^{\frac{1}{2}} a^{(1)}_{r2} + (r+1)^{\frac{1}{2}} a^{(1)}_{r+1,2}\right]. \tag{4.32d}$$

We shall discuss the solutions of these equations in the next two sections.

Going on to the third order equation (4.20d), we put in the right hand side $h^{(0)}_{\omega,k}$ given by (4.25a), $h^{(1)}_{\omega,k}$ given by (4.27), and $h^{(2)}_{\omega,k}$ given by (4.31). Requiring that the right hand side be orthogonal to φ_1, φ_2 and φ_3 gives three equations determining $E^{(3)}$, $A^{(2)}_2$ and $A^{(2)}_3$:

$$E^{(3)} = -(\tfrac{1}{6})^{\frac{1}{2}} a^{(2)}_{11} + (\tfrac{1}{3})^{\frac{1}{2}} a^{(2)}_{02},$$

$$A^{(2)}_2 = (\tfrac{3}{5})^{\frac{1}{2}} a^{(2)}_{11} - (\tfrac{6}{5})^{\frac{1}{2}} E^{(2)} A^{(1)}_2$$

$$= (\tfrac{3}{5})^{\frac{1}{2}} a^{(2)}_{11} + (\tfrac{3}{25})^{\frac{1}{2}} (a^{(1)}_{11})^2 + (\tfrac{6}{25})^{\frac{1}{2}} a^{(1)}_{11} a^{(1)}_{02},$$

$$A_3^{(2)} = -\left(\frac{1}{20}\right)^{\frac{1}{2}} a_{11}^{(2)} - \left(\frac{1}{10}\right)^{\frac{1}{2}} a_{02}^{(2)} - \left(\tfrac{3}{10}\right)^{\frac{1}{2}} E^{(2)} A_3^{(1)} \tag{4.33}$$

$$= -\left(\frac{1}{20}\right)^{\frac{1}{2}} a_{11}^{(2)} - \left(\frac{1}{10}\right)^{\frac{1}{2}} a_{02}^{(2)} + \tfrac{1}{10}\left(a_{02}^{(1)}\right)^2 - \tfrac{1}{40}\left(a_{11}^{(1)}\right)^2 .$$

Here we shall stop, having determined the sound normal mode solution through second order and the corresponding eigenvalue E through third order.

But we get something more for free, at least in principle. This stems from the fact that in perturbation theory for self-adjoint operators the eigenvalue through $2n$-th and $(2n+1)$-th order is determined by the eigenfunction through n-th order [6]. To see how this works, consider the example of the third order eigenvalue which can be expressed in terms of the first order eigenfunction as follows. Form the scalar product of the third order equation (4.20d) with $h_{\omega,k}^{(0)}$. Since $\boldsymbol{J}$ is self-adjoint, the left hand side vanishes, and since by construction $h_{\omega,k}^{(0)}$ is normalized to unity and orthogonal to $h_{\omega,k}^{(1)}$, we get

$$E^{(3)} = -\left(h_{\omega,k}^{(0)}, \left[E^{(1)} - c_z\right] h_{\omega,k}^{(2)}\right) . \tag{4.34}$$

If we put $h_{\omega,k}^{(2)}$ given by (4.31) and $h_{\omega,k}^{(0)}$ given by (4.25a) in this equation and use (4.23a), we get our previous result (4.33). Instead we note that:

$$\left(h_{\omega,k}^{(0)}, \left[E^{(1)} - c_z\right] h_{\omega,k}^{(2)}\right) = \left(\left[E^{(1)} - c_z\right] h_{\omega,k}^{(0)}, h_{\omega,k}^{(2)}\right)$$

$$= \left(\boldsymbol{J} h_{\omega,k}^{(1)}, h_{\omega,k}^{(2)}\right)$$

$$= \left(h_{\omega,k}^{(1)}, \boldsymbol{J} h_{\omega,k}^{(2)}\right) . \tag{4.35}$$

Here in the first step we used the self-adjoint character of $(E^{(1)} - c_z)$, in the second step we used the first order equation (4.20b), and in the third step we used the self-adjoint character of $\boldsymbol{J}$. Finally, using the second order equation (4.20c) and the fact that $h_{\omega,k}^{(1)}$ is orthogonal to $h_{\omega,k}^{(0)}$, we obtain the desired result expressing $E^{(3)}$ in terms of the first order solutions:

$$E^{(3)} = -\left(h_{\omega,k}^{(1)}, \left[E^{(1)} - c_z\right] h_{\omega,k}^{(1)}\right) . \tag{4.36}$$

By similar manipulations of the fourth and fifth order equations we obtain the following expressions for $E^{(4)}$ and $E^{(5)}$ in terms of the normal mode solutions through second order [7]:

$$E^{(4)} = -\left(h^{(1)}_{\omega,k}, \left[E^{(1)} - c_z\right] h^{(2)}_{\omega,k}\right) - E^{(2)}\left(h^{(1)}_{\omega,k}, h^{(1)}_{\omega,k}\right), \tag{4.37}$$

$$E^{(5)} = -\left(h^{(2)}_{\omega,k}, \left[E^{(1)} - c_z\right] h^{(2)}_{\omega,k}\right)$$
$$- E^{(2)}\left[\left(h^{(1)}_{\omega,k}, h^{(2)}_{\omega,k}\right) + \left(h^{(2)}_{\omega,k}, h^{(1)}_{\omega,k}\right)\right] - E^{(3)}\left(h^{(1)}_{\omega,k}, h^{(1)}_{\omega,k}\right). \tag{4.38}$$

Thus, in this section we have formally determined E through fifth order in ϵ, which corresponds to an expansion of the complex frequency of the sound mode in powers of the wave number through the fifth power. In the next two sections we discuss the numerical determination of the coefficients, first in the case of Maxwell molecules and then in the general case.

4. SOLUTION OF THE EQUATIONS FOR MAXWELL MOLECULES

As we saw in chapter III, the Burnett functions are eigenfunctions of the collision operator for the special case of Maxwell molecules, i.e., molecules which repel each other with a force proportional to the inverse fifth power of their separation. In this case the equations to be solved in the perturbation method outlined in the previous section become essentially trivial. Thus, for Maxwell molecules,

$$J^{l}_{rr'} = \lambda_{rl}\, \delta_{rr'}, \tag{4.39}$$

where λ_{rl} is given by (3.95). Putting this expression in equations (4.28) we find immediately that

$$a^{(1)}_{11} = \frac{1}{6^{\frac{1}{2}} \lambda_{11}}, \qquad a^{(1)}_{02} = -\frac{1}{3^{\frac{1}{2}} \lambda_{02}}, \tag{4.40}$$

and all other $a^{(1)}_{rl}$ vanish. Using this result in (4.29) we find

$$E^{(2)} = -\frac{1}{6\lambda_{11}} - \frac{1}{3\lambda_{02}}. \tag{4.41}$$

In the same way, using the result (4.40) in the right hand side of the second order equations (4.32), we find that

$$a_{20}^{(2)} = \frac{1}{3\lambda_{11}\lambda_{20}}, \qquad a_{11}^{(2)} = \left(\frac{1}{80}\right)^{\frac{1}{2}} \frac{1}{\lambda_{11}^2} - \left(\frac{1}{20}\right)^{\frac{1}{2}} \frac{1}{\lambda_{11}\lambda_{02}},$$

$$a_{02}^{(2)} = \left(\frac{1}{40}\right)^{\frac{1}{2}} \frac{1}{\lambda_{11}\lambda_{02}} - \left(\frac{8}{45}\right)^{\frac{1}{2}} \frac{1}{\lambda_{02}^2},$$

$$a_{12}^{(2)} = -\left(\frac{7}{45}\right)^{\frac{1}{2}} \frac{1}{\lambda_{11}\lambda_{12}}, \qquad a_{03}^{(2)} = \left(\frac{3}{10}\right)^{\frac{1}{2}} \frac{1}{\lambda_{02}\lambda_{03}}, \tag{4.42}$$

and all other $a_{rl}^{(2)}$ vanish. Using these results in (4.33) we find

$$E^{(3)} = -\left(\tfrac{5}{6}\right)^{\frac{1}{2}} \left[\frac{1}{20\lambda_{11}^2} - \frac{1}{5\lambda_{11}\lambda_{02}} + \frac{1}{15\lambda_{02}^2}\right]. \tag{4.43}$$

Finally, using these results in (4.37) gives:

$$E^{(4)} = -\frac{1}{18\lambda_{02}^3} + \frac{2}{15\lambda_{02}^2\lambda_{11}} + \frac{1}{9\lambda_{02}\lambda_{11}^2} + \frac{1}{30\lambda_{11}^3} - \frac{3}{10\lambda_{02}^2\lambda_{03}} - \frac{1}{9\lambda_{11}^2\lambda_{20}} - \frac{7}{45\lambda_{11}^2\lambda_{12}}, \tag{4.44}$$

and in (4.38) gives:

$$E^{(5)} = \left(\tfrac{5}{6}\right)^{\frac{1}{2}} \left[\frac{23}{150\lambda_{02}^4} - \frac{1}{30\lambda_{02}^3\lambda_{11}} - \frac{1}{450\lambda_{02}^2\lambda_{11}^2} - \frac{13}{300\lambda_{02}\lambda_{11}^3} + \frac{59}{1440\lambda_{11}^4} - \frac{12}{25\lambda_{02}^3\lambda_{03}} + \frac{9}{50\lambda_{02}^2\lambda_{11}\lambda_{03}} + \frac{2}{15\lambda_{02}\lambda_{11}^2\lambda_{20}} + \frac{14}{75\lambda_{02}\lambda_{11}^2\lambda_{12}}\right.$$

$$+\frac{6}{25\lambda_{02}\lambda_{11}\lambda_{03}\lambda_{12}}-\frac{3}{10\lambda_{02}^{2}\lambda_{03}^{2}}-\frac{1}{15\lambda_{11}^{3}\lambda_{20}}$$
$$-\frac{7}{75\lambda_{11}^{3}\lambda_{12}}-\frac{1}{9\lambda_{11}^{2}\lambda_{20}^{2}}-\frac{7}{45\lambda_{11}^{2}\lambda_{12}^{2}}\Big]. \qquad (4.45)$$

These formulas for the perturbed eigenvalues can be simplified considerably if we use the fact that the eigenvalues involved are simple rational multiples of one another. Thus, from (3.95) we can readily verify that

$$\lambda_{11}=\tfrac{2}{3}\lambda_{02}, \qquad \lambda_{03}=\tfrac{3}{2}\lambda_{02}, \qquad \lambda_{20}=\tfrac{2}{3}\lambda_{02}, \qquad \lambda_{12}=\tfrac{7}{6}\lambda_{02}\ , \qquad (4.46)$$

and, therefore,

$$E^{(1)}=(\tfrac{5}{6})^{\frac{1}{2}}\ ,$$
$$E^{(2)}=-\tfrac{7}{12}\frac{1}{\lambda_{02}}\ ,$$
$$E^{(3)}=-(\tfrac{5}{6})^{\frac{1}{2}}\tfrac{19}{240}\frac{1}{\lambda_{02}^{2}}\ ,$$
$$E^{(4)}=-\tfrac{53}{144}\frac{1}{\lambda_{02}^{3}}\ ,$$
$$E^{(5)}=-(\tfrac{5}{6})^{\frac{1}{2}}\tfrac{47,063}{89,600}\frac{1}{\lambda_{02}^{4}}\ . \qquad (4.47)$$

Just as in chapter II, we now want to express the dispersion law as an expansion of the dimensionless frequency ξ in terms of the dimensionless wave number x. Referring to (2.36), we have

$$\xi=\frac{3\eta}{5nkT}\,\omega\ , \qquad x=\frac{\eta}{nm}\left(\frac{3m}{5kT}\right)^{\frac{1}{2}}k\ . \qquad (4.48)$$

But the expression for the viscosity coefficient η in a gas of Maxwell molecules involves the eigenvalue λ_{02} [8]:

$$\eta=-\frac{m}{2\sigma\lambda_{02}}\left(\frac{2kT}{m}\right)^{\frac{1}{2}}\ . \qquad (4.49)$$

Hence, if we use (4.5) to relate E to ξ and ϵ to x, we can express the expansion (4.19) with the coefficients given by (4.47), in the form:

$$\xi = x - \frac{7}{6} i\, x^2 + \frac{19}{72} x^3 + \frac{265}{108}\, i\, x^4 - \frac{47\,063}{8064}\, x^5 + \dots\ . \tag{4.50}$$

This expansion is to be compared with the corresponding expansion (2.50) obtained from the hydrodynamic equations.

In comparing these results with experiments, in which the phase velocity and absorption length are measured as functions of frequency, it is necessary to invert the series (4.50) to obtain the expansion:

$$x = \xi + \frac{7}{6} i \xi^2 - \frac{215}{72}\, \xi^3 - \frac{5155}{432} i \xi^4 + \frac{4\,115\,101}{72\,576}\, \xi^5 + \dots\ . \tag{4.51}$$

Since, for real ω, the phase velocity and absorption length are given by (1.3) and (1.4), then referring to (4.49) we can write

$$\frac{v_0}{v_{ph}} = \mathrm{Re}\left\{\frac{x}{\xi}\right\} = 1 - \frac{215}{72}\, \xi^2 + \frac{4\,115\,101}{72\,576}\, \xi^4 + \dots\ , \tag{4.52}$$

$$\frac{\alpha v_0}{\omega} = \mathrm{Im}\left\{\frac{x}{\xi}\right\} = \frac{7}{6}\, \xi - \frac{5155}{432}\, \xi^3 + \dots\ . \tag{4.52}$$

Here $v_0 = (5kT/3m)^{\frac{1}{2}}$ is the Laplace expression for the sound speed at low frequencies.

The coefficients in the expansion (4.52) are of course identical with those found by Chang and Uhlenbeck using the determinental method. Note, incidentally, that the coefficients of the higher powers of ξ will no longer be rational numbers, since the higher λ_{rl} are in general not rational multiples of λ_{02}.

5. APPROXIMATE SOLUTION OF THE EQUATIONS FOR ARBITRARY POTENTIALS

For potentials other than that of Maxwell molecules the equations (4.28) and corresponding equations occurring in higher order cannot be solved in closed form. However, these same equations occur in the Chapman-Enskog development of the hydrodynamic equations from the Boltzmann equation and there it was shown long ago by Chapman, Cowling and Burnett that a solution by successive truncations coverges very rapidly [9]. In this section we apply their method to obtain approximate solutions for the equations appearing in the perturbation expansion.

To see how the method works, consider first the equations (4.28a). In the first approximation we include only as many terms in the sum on the left hand side as appear in the inhomogeneous term on the right hand side. That is, we approximate the infinite sequence of coupled equations (4.28a) by the single equation:

$$J^1_{11}\left[a^{(1)}_{11}\right]_1 = (\tfrac{1}{6})^{\frac{1}{2}} . \tag{4.53}$$

Here we have adopted the convention of Chapman and Cowling in which successive approximations are indicated by a square bracket and a subscript. The first approximate solution is therefore:

$$\left[a^{(1)}_{11}\right]_1 = (\tfrac{1}{6})^{\frac{1}{2}} \frac{1}{J^1_{11}} , \tag{4.54}$$

and all other $[a^{(1)}_{r1}]_1$ vanish. Note that this first approximation is exact in the case of Maxwell molecules.

In the second approximation we include one more term on the left hand side, i.e., we replace the equation (4.28a) with the pair of coupled equations:

$$J^1_{11}\left[a^{(1)}_{11}\right]_2 + J^1_{12}\left[a^{(1)}_{21}\right]_2 = (\tfrac{1}{6})^{\frac{1}{2}} ,$$

$$J^1_{12}\left[a^{(1)}_{11}\right]_2 + J^1_{22}\left[a^{(1)}_{21}\right]_2 = 0 . \tag{4.55}$$

Hence, the second approximate solution is

$$\left[a^{(1)}_{11}\right]_2 = (\tfrac{1}{6})^{\frac{1}{2}} \frac{J^1_{22}}{J^1_{11}J^1_{22} - (J^1_{21})^2} ,$$

$$\left[a^{(1)}_{21}\right]_2 = -(\tfrac{1}{6})^{\frac{1}{2}} \frac{J^1_{12}}{J^1_{11}J^1_{22} - (J^1_{21})^2} , \tag{4.56}$$

and all other $[a^{(1)}_{r1}]_2$ vanish. Continuing in this way, adding in each successive approximation one more term on the left hand side, we obtain a sequence of successive approximations which in practice converges rapidly [10].

In exactly the same way we find from the equations (4.28b) the first approximate solution:

$$\left[a_{02}^{(1)}\right]_1 = -\left(\tfrac{1}{3}\right)^{\frac{1}{2}} \frac{1}{J_{00}^{2}} , \tag{4.57}$$

and the second approximate solution:

$$\left[a_{02}^{(1)}\right]_2 = -\left(\tfrac{1}{3}\right)^{\frac{1}{2}} \frac{J_{11}^{2}}{J_{00}^{2} J_{11}^{2} - (J_{01}^{2})^2} ,$$

$$\left[a_{12}^{(1)}\right]_2 = \left(\tfrac{1}{3}\right)^{\frac{1}{2}} \frac{J_{01}^{2}}{J_{00}^{2} J_{11}^{2} - (J_{01}^{2})^2} . \tag{4.58}$$

Turning next to the second order equations (4.32), in the first approximation we put in the right hand side the first approximations (4.54) and (4.57) to the first order equations and keep in each case only as many terms in the sum occurring on the left as have a non-vanishing inhomogeneous term. Thus, the first approximation to (4.32a) is

$$J_{22}^{0} \left[a_{20}^{(2)}\right]_1 = \left(\tfrac{2}{3}\right)^{\frac{1}{2}} \left[a_{11}^{(1)}\right]_1 , \tag{4.59}$$

the first approximation to (4.32b) is

$$J_{11}^{1} \left[a_{11}^{(2)}\right]_1 = \left(\tfrac{3}{40}\right)^{\frac{1}{2}} \left[a_{11}^{(1)}\right]_1 + \left(\tfrac{3}{20}\right)^{\frac{1}{2}} \left[a_{02}^{(1)}\right]_1 , \tag{4.60}$$

but the first approximation to (4.32c) is the pair of coupled equations:

$$J_{00}^{2} \left[a_{02}^{(2)}\right]_1 + J_{01}^{2} \left[a_{12}^{(2)}\right]_1 = \left(\tfrac{3}{20}\right)^{\frac{1}{2}} \left[a_{11}^{(1)}\right]_1 + \left(\tfrac{8}{15}\right)^{\frac{1}{2}} \left[a_{02}^{(1)}\right]_1 ,$$

$$J_{01}^{2} \left[a_{02}^{(2)}\right]_1 + J_{11}^{2} \left[a_{12}^{(2)}\right]_1 = -\left(\tfrac{14}{15}\right)^{\frac{1}{2}} \left[a_{11}^{(1)}\right]_1 , \tag{4.61}$$

while the first approximation to (4.32d) is again the single equation:

$$J_{00}^{3} \left[a_{03}^{(2)}\right]_1 = -\left(\tfrac{9}{10}\right)^{\frac{1}{2}} \left[a_{02}^{(1)}\right]_1 . \tag{4.62}$$

Using (4.54) and (4.57) and solving these equations we find

$$\left[a_{20}^{(2)}\right]_1 = \frac{1}{3}\frac{1}{J_{11}^1 J_{22}^0}$$

$$\left[a_{11}^{(2)}\right]_1 = \left(\frac{1}{80}\right)^{\frac{1}{2}} \frac{1}{(J_{11}^1)^2} - \left(\frac{1}{20}\right)^{\frac{1}{2}} \frac{1}{J_{11}^1 J_{00}^2}$$

$$\left[a_{02}^{(2)}\right]_1 = \frac{\left(\frac{1}{40}\right)^{\frac{1}{2}} J_{00}^2 J_{11}^2 - \left(\frac{8}{45}\right)^{\frac{1}{2}} J_{11}^1 J_{11}^2 + \left(\frac{7}{45}\right)^{\frac{1}{2}} J_{01}^2 J_{00}^2}{J_{11}^1 J_{00}^2 \left[J_{00}^2 J_{11}^2 - (J_{01}^2)^2\right]}$$

$$\left[a_{12}^{(2)}\right]_1 = \frac{-\left(\frac{7}{45}\right)^{\frac{1}{2}} (J_{00}^2)^2 - \left(\frac{1}{40}\right)^{\frac{1}{2}} J_{01}^2 J_{00}^2 + \left(\frac{8}{45}\right)^{\frac{1}{2}} J_{01}^2 J_{11}^1}{J_{11}^1 J_{00}^2 \left[J_{00}^2 J_{11}^2 - (J_{01}^2)^2\right]}$$

$$\left[a_{03}^{(2)}\right]_1 = \left(\frac{9}{20}\right)^{\frac{1}{2}} \frac{1}{J_{00}^2 J_{00}^3} . \tag{4.63}$$

With these results we get from (4.25), (4.29) and (4.33) that:

$$E^{(1)} = \left(\tfrac{5}{6}\right)^{\frac{1}{2}} , \qquad [E^{(2)}]_1 = -\frac{1}{6J_{11}^1} - \frac{1}{3J_{00}^2} ,$$

$$[E^{(3)}]_1 = \left(\tfrac{5}{6}\right)^{\frac{1}{2}} \left[\left\{-J_{11}^2 \left[3(J_{00}^2)^2 - 12 J_{11}^1 J_{00}^2 + 16 (J_{11}^1)^2\right] + 3(J_{00}^2 - 2J_{11}^1)(J_{01}^2)^2 + 4\times 14^{\frac{1}{2}} J_{01}^2 J_{11}^1 J_{00}^2\right\} \times \left\{60 J_{00}^2 (J_{11}^1)^2 \left[J_{00}^2 J_{11}^2 - (J_{01}^2)^2\right]\right\}^{-1}\right] . \tag{4.64}$$

At this point one might think that, as in the case of Maxwell molecules, we could determine $[E^{(4)}]_1$, and $[E^{(5)}]_1$, through the identities (4.37) and (4.38). However, these identities, while they are correct for the exact solutions, do not hold for the approximate solutions. That is, if one puts the first approximate solutions (4.54), (4.57) and (4.63) in the right hand side of (4.37) the result is not the first approximate expression for $E^{(4)}$ which we would obtain by putting these same first approximate solutions in the right hand side of the third order equation (4.20d), solving that equation in first approximation, and then determining $[E^{(4)}]_1$ from the fourth order equation 11). The reason for this discrepancy is that in our approximation scheme the truncations of the collision operator are not the same in every order of the perturbation expansion, so the manipulations used in deriving (4.37) and (4.38) are not valid. Therefore, to be consistent one must solve in first approximation the third order equations to find $[E^{(4)}]_1$, and the fourth order equations to find $[E^{(5)}]_1$.

These calculations are obviously straightforward but very tedious. We have calculated $[E^{(4)}]_1$, for which we find the general expression:

$$[E^{(4)}]_1 = \frac{1}{J^1_{11}J^1_{22} - (J^1_{12})^2} \times$$

$$\Bigg\{ -J^1_{22}\Bigg[\frac{56\,(J^2_{00})^2 + 3J^2_{11}(3J^2_{00} - 8J^1_{11}) + 4\times 14^{\frac{1}{2}}J^2_{01}\,(3J^2_{00} - 4J^1_{11})}{360\,J^1_{11}J^2_{00}\,[J^2_{00}J^2_{11} - (J^2_{01})^2]}$$

$$- \frac{1}{30(J^1_{11})^2} - \frac{41}{360J^1_{11}J^2_{00}} + \frac{1}{180\,(J^2_{00})^2} + \frac{1}{9J^1_{11}J^0_{22}}\Bigg]$$

$$- 7^{\frac{1}{2}}J^1_{12}\Bigg[\frac{28(J^2_{00})^2 + 14^{\frac{1}{2}}J^2_{01}(3J^2_{00} - 8J^1_{11})}{630J^1_{11}J^2_{00}\,[J^2_{00}J^2_{11} - (J^2_{01})^2]} + \frac{1}{18J^1_{11}J^0_{22}}\Bigg]\Bigg\}$$

$$+ \frac{1}{J^2_{00}J^2_{11} - (J^2_{01})^2}\Bigg\{J^2_{11}\Bigg[\frac{-J^2_{11}(3J^2_{00} - 8J^1_{11}) + 2\times 14^{\frac{1}{2}}J^2_{01}J^2_{00}}{45J^1_{11}J^2_{00}\,[J^2_{00}J^2_{11} - (J^2_{01})^2]}$$

$$+ \frac{1}{45(J^1_{11})^2} + \frac{1}{180J^1_{11}J^2_{00}} + \frac{11}{90(J^2_{00})^2} - \frac{3}{10J^2_{00}J^3_{00}}\Bigg]$$

$$+ 14^{\frac{1}{2}}J^2_{01}\Bigg[\frac{-28(J^2_{00})^2 + 14^{\frac{1}{2}}J^2_{01}\,(3J^2_{00} - 8J^1_{11})}{504J^1_{11}J^2_{00}\,[J^2_{00}J^2_{11} - (J^2_{01})^2]}$$

$$- \frac{3}{70 J^2_{00} J^3_{00}} + \frac{1}{60 (J^1_{11})^2} - \frac{1}{30 J^1_{11} J^2_{00}} \Big]\Big\} \ . \qquad (4.65)$$

The corresponding expression for $E^{(5)}$ in first approximation has not been obtained. As we shall see, the experimental results are not yet sufficiently accurate to afford a meaningful comparison with this coefficient.

The matrix elements which appear in the expressions (4.64) and (4.65) are not all independent. In fact, from appendix A we can readily verify that we can write:

$$J^0_{22} = \tfrac{2}{3} J^2_{00} , \quad J^1_{11} = \tfrac{2}{3} J^2_{00} , \quad J^3_{00} = \tfrac{3}{2} J^2_{00} ,$$

$$J^1_{12} = - \frac{\delta}{63^{\frac{1}{2}}} J^2_{00} , \quad J^2_{01} = - \frac{\delta}{56^{\frac{1}{2}}} J^2_{00} ,$$

$$J^1_{22} = (1 + \tfrac{1}{14}\gamma) J^2_{00} , \quad J^2_{11} = \tfrac{7}{6} (1 + \tfrac{9}{196}\gamma) J^2_{00} , \qquad (4.66)$$

where

$$\gamma = \frac{35 \Omega^{(2)}(2) - 28 \Omega^{(2)}(3) + 4 \Omega^{(2)}(4)}{3 \Omega^{(2)}(2)} ,$$

$$\delta = \frac{-7 \Omega^{(2)}(2) + 2 \Omega^{(2)}(3)}{\Omega^{(2)}(2)} . \qquad (4.67)$$

For the case of Maxwell molecules the quantities γ and δ vanish and we recover the identities (4.46) between the eigenvalues.

Using the identities (4.66) in the expressions (4.64) and (4.65), we can write

$$E^{(1)}_1 = (\tfrac{5}{6})^{\frac{1}{2}} , \qquad [E^{(2)}]_1 = - \tfrac{7}{12} \frac{1}{J^2_{00}} ,$$

$$[E^{(3)}]_1 = - (\tfrac{5}{6})^{\frac{1}{2}} \tfrac{19}{240} (1 + \Delta_1) \frac{1}{(J^2_{00})^2} ,$$

$$[E^{(4)}]_1 = - \tfrac{53}{144} (1 + \Delta_2) \frac{1}{(J^2_{00})^3} , \qquad (4.68)$$

where

$$\Delta_1 = \frac{84\delta(24+\delta)}{19[196+3(3\gamma-\delta^2)]},$$

$$\Delta_2 = -\frac{3}{530[196+3(3\gamma-\delta^2)][42+3\gamma-\delta^2]}[54\,432\,\gamma + 373\,968\,\delta$$

$$+ 3888\,\gamma^2 + 12\,348\,\gamma\delta - 36\,274\,\delta^2 - 2277\,\gamma\delta^2 - 3780\,\delta^3 + 390\,\delta^4]$$

$$+ \frac{18}{265\,[196+3(3\gamma-\delta^2)]^2}[63\,896\,\delta + 963\,\gamma\delta + 6076\,\delta^2 + 414\,\delta^3 - 35\,\delta^4]\,. \tag{4.69}$$

6. THE DISPERSION LAW FOR POWER LAW AND LENNARD-JONES MODELS

Here, as in chapter II and in our discussion of the Maxwell model, we express the dispersion law for sound as a relation between the dimensionless frequency variable ξ and the dimensionless wave number variable x, defined in (4.48). But in the Chapman-Enskog development [12)] of the hydrodynamic equations the viscosity η, and also the thermal conductivity κ, are expressed in terms of the solutions of equations which are identical with the first order equations (4.28). In fact

$$\eta = \frac{m}{\sigma}\left(\frac{3\,kT}{2m}\right)^{\frac{1}{2}} a_{02}^{(1)}\,, \qquad \kappa = -\frac{5k}{2\sigma}\left(\frac{3\,kT}{m}\right)^{\frac{1}{2}} a_{11}^{(1)}\,. \tag{4.70}$$

Hence, putting this expression for η in the definitions (4.48) and then eliminating ω and k between these equations and the definitions (4.5) of E and ϵ, we find

$$E = -i\,\frac{5}{3\times3^{\frac{1}{2}}\,a_{02}^{(1)}}\,\xi\,, \qquad \epsilon = -i\,\frac{10^{\frac{1}{2}}}{3\,a_{02}^{(1)}}\,x\,. \tag{4.71}$$

The expansion (4.19) of E in powers of ϵ becomes, therefore, the expansion:

$$\xi = \sum_{n=1}^{\infty} (-i)^{n-1} e^{(n)} x^n$$

$$= x - i\, e^{(2)} x^2 - e^{(3)} x^3 + i\, e^{(4)} x^4 + \dots\ , \tag{4.72}$$

where

$$e^{(n)} = \left(\tfrac{5}{6}\right)^{\frac{1}{2}n-1} \left[\frac{2}{3^{\frac{1}{2}} a_{02}^{(1)}}\right]^{n-1} E^{(n)}\ . \tag{4.73}$$

In the case of power law forces the coefficients $e^{(n)}$ are purely numerical quantities, dependent only upon the power law index s. This follows from the form of the collision cross section for power law forces [see (3.82)] which implies that the linearized collision operator in this case is of the form:

$$\boldsymbol{J} = \frac{a^2}{\sigma} \left(\frac{\varphi_s}{kT}\right)^{\frac{2}{s}} \boldsymbol{j}\ , \tag{4.74}$$

where $\boldsymbol{j}$ is an operator depending only upon s, independent of the temperature and of the strength of the force. Therefore, by inspection of the equations (4.20) of the perturbation expansion we see that the quantities:

$$\left[\frac{a^2}{\sigma} \left(\frac{\varphi_s}{kT}\right)^{\frac{2}{s}}\right]^n h_{\omega,k}^{(n)}\ , \qquad \left[\frac{a^2}{\sigma} \left(\frac{\varphi_s}{kT}\right)^{\frac{2}{s}}\right]^{n-1} E^{(n)} \tag{4.75}$$

are dependent only upon s. Hence, since $a_{02}^{(1)} = (\psi_{02}, h_{\omega,k}^{(1)})$, the factors in $E^{(n)}$ and $a_{02}^{(1)}$ cancel in (4.73), leaving only a dependence upon s.

By a simple extension of this argument we can show that for the Lennard-Jones model, in which the intermolecular potential is of the form:

$$\varphi(r) = 4\, \epsilon_0 \left[\left(\frac{a}{r}\right)^{12} - \left(\frac{a}{r}\right)^{6}\right], \tag{4.76}$$

the coefficients $e^{(n)}$ depend only upon the so-called reduced temperatur kT/ϵ_0.

Finally, we remark that the coefficients $e^{(n)}$ in the expansion (4.72) are identical with the coefficients in the corresponding expansion obtained from the successive hydrodynamic equations, as outlined in chapter II. For $e^{(2)}$ this is simple to see since, putting (4.29) in (4.72) for $n = 2$, we have

$$e^{(2)} = \tfrac{2}{3} - \frac{2^{\frac{1}{2}} a_{11}^{(1)}}{3a_{02}^{(1)}} . \tag{4.77}$$

But, from (4.70) we see that

$$\frac{2^{\frac{1}{2}} a_{11}^{(1)}}{a_{02}^{(1)}} = -\tfrac{2}{5} \frac{m\kappa}{k\eta} = -\tfrac{3}{5} f , \tag{4.78}$$

where f is the Eucken number (2.33). Hence,

$$e^{(2)} = \tfrac{2}{3} + \tfrac{1}{5} f , \tag{4.79}$$

which is exactly the coefficient of the corresponding term in the expansion (2.50) obtained from the hydrodynamic equations. The demonstration of the identity of $e^{(3)}$ with the coefficient of the next term in the expansion (2.48) is only more complicated. It hinges on the identity of the equations to be solved in second approximation of the Chapman-Enskog scheme with our second order equations (4.30). Since, so far as we know, the Chapman-Enskog scheme has never been carried beyond the second approximation, we can make no comparison with the higher coefficients, but it seems safe enough to assume that the approach via Chapman-Enskog and the hydrodynamic equations and our perturbation approach yield identical results.

If we put the first approximate results (4.57) and (4.68) in (4.73), we get the following first approximate expressions for the coefficients in the expansion (4.72):

$$[e^{(2)}]_1 = \tfrac{7}{6} ,$$

$$[e^{(3)}]_1 = -\tfrac{19}{72} (1 + \Delta_1) ,$$

$$[e^{(4)}]_1 = \tfrac{265}{108} (1 + \Delta_2) . \tag{4.80}$$

Here Δ_1 and Δ_2 are given by (4.69).

For the power law model the quantities Δ_1 and Δ_2 can be expressed in terms of the power law index s:

$$\Delta_1 = \frac{42(s-4)(25s-4)}{19s(101s-12)}, \tag{4.81a}$$

$$\Delta_2 = -\frac{21(s-4)(3\,120\,757\,s^4 + 1\,415\,648\,s^3 - 479\,424\,s^2 - 93\,696\,s + 22\,272)}{4240\,s^2(101\,s-12)^2(11\,s-2)}. \tag{4.81b}$$

Both Δ_1 and Δ_2 vanish for Maxwell molecules (s=4) and we recover the expansion (4.50). As s increases Δ_1 increases while Δ_2 decreases until at $s = \infty$ (hard spheres) they have the values:

$$\Delta_1 = \frac{1050}{1919} \approx 0.5472 ,$$

$$\Delta_2 = -\frac{65\,535\,897}{475\,774\,640} \approx -0.1377 . \tag{4.82}$$

For the Lennard-Jones model the Ω-integrals which appear in (4.67) have been tabulated as functions of the reduced temperature kT/ϵ_0 by Hirschfelder, Curtis and Bird [13] (for use in calculation of transport coefficients). It is a simple matter (with the aid of a computer) to use their tables to construct the tables of Δ_1 and Δ_2 given in appendix B. In fig. 5 we show Δ_1 plotted as a function of reduced temperature. The largest value of Δ_1 occurs in the high temperature limit, where Δ_1 is not very different from its value for an inverse power potential with $s = 12$. For lower temperatures Δ_1 decreases, the slight dip near $kT/\epsilon_0 = 2$ is apparently the vestige of orbitting collisions which occur because of the attractive part of the Lennard-Jones potential.

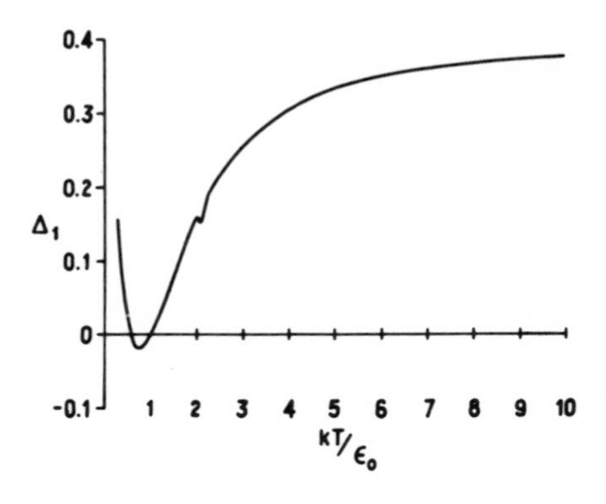

Fig. 5. The quantity Δ_1 as a function of reduced temperature.

7. COMPARISON OF THE FIRST APPROXIMATE RESULTS WITH EXPERIMENT

With the first approximate results (4.80) for the coefficients, the expansion (4.72) of the dimensionless frequency ξ in powers of the dimensionless wave number x becomes:

$$\xi = x - \frac{7}{6} i\, x^2 + \frac{19}{72} (1 + \Delta_1) x^3 + \frac{265}{108}\, i (1 + \Delta_2) x^4 + \dots , \quad (4.83)$$

where Δ_1 and Δ_2, given by (4.69) measure the deviation of the coefficients from their values for Maxwell molecules, for which Δ_1 and Δ_2 vanish. These coefficients depend significantly upon the molecular model as evidenced by the example of hard spheres, for which $\Delta_1 \approx 0.55$, $\Delta_2 \approx -0.14$.

But for comparison with experimental measurements of sound dispersion in monoatomic gases, in which the phase velocity and absorption length are measured as functions of frequency, it is necessary to invert the series (4.83) to express the dimensionless wave number in powers of the dimensionless frequency. The result is

$$x = \xi + \frac{7}{6} i\, \xi^2 - \frac{215}{72} \left(1 + \frac{19\Delta_1}{215}\right) \xi^3$$
$$- \frac{5155}{432} i \left(1 + \frac{133\,\Delta_1 + 212\,\Delta_2}{1031}\right) \xi^4 + \dots . \quad (4.84)$$

Here it is striking that the coefficients in this expansion depend much more weakly on the molecular model than do those in the expansion (4.83). This is due to a sort of numerical accident by which Δ_1 and Δ_2 are multiplied by small numerical factors. Thus, for the case of hard spheres the coefficient of ξ^3 in (4.84) differs by only about 5% from its Maxwell value while the corresponding difference in the coefficient of x^3 in (4.83) is 55%. A similar phenomenon in the coefficient of ξ^4 is enhanced by the partial cancellation due to the opposite signs of Δ_1 and Δ_2.

Referring to (4.48) and the expressions (1.3) and (1.4) for the phase velocity and the absorption length for real frequencies, we can write

$$\left[\frac{v_0}{v_{ph}}\right]_1 = \text{Re}\left\{\frac{x}{\xi}\right\} = 1 - \frac{215}{72}\left(1 + \frac{19}{215}\Delta_1\right)\xi^2 + \dots , \quad (4.85a)$$

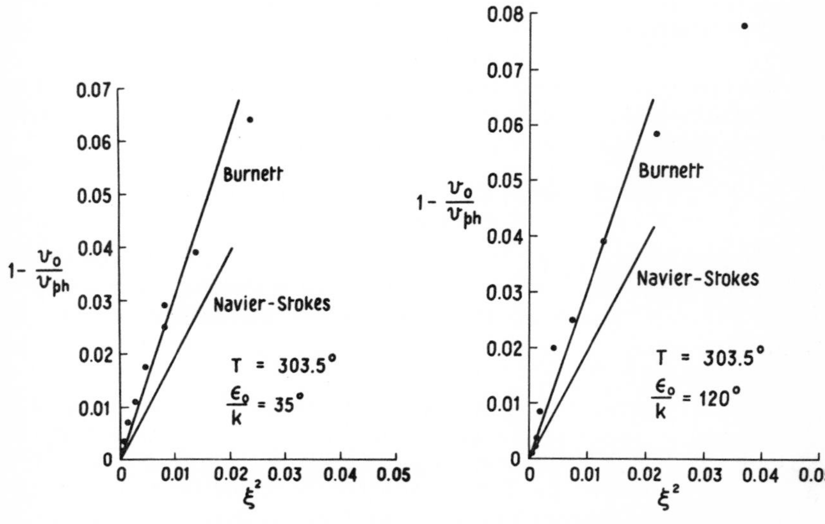

Fig. 6. The dispersion of sound for neon.

Fig. 7. The dispersion of sound for argon.

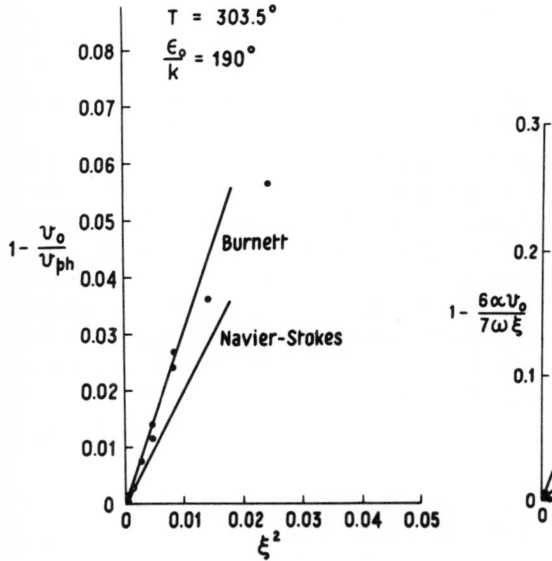

Fig. 8. The dispersion of sound for krypton.

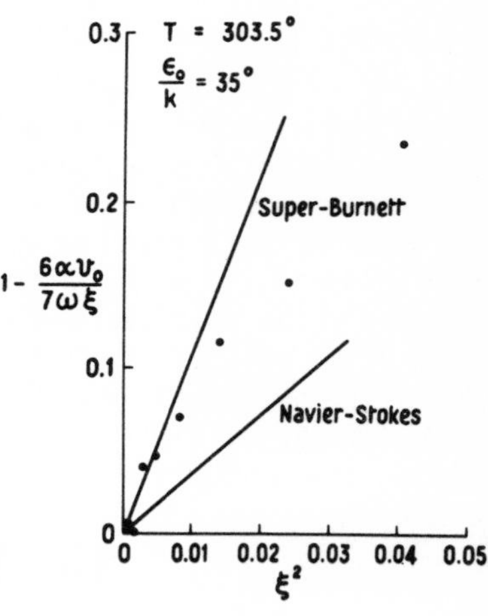

Fig. 9. The absorption of sound for neon.

$$\left[\frac{\alpha v_0}{\omega}\right]_1 = \mathrm{Im}\left\{\frac{x}{\xi}\right\} = \frac{7}{6}\xi - \frac{5155}{432}\left(1 + \frac{133\,\Delta_1 + 212\,\Delta_2}{1031}\right)\xi^3 + \dots \quad , \qquad (4.85b)$$

where $v_0 = (5kT/3m)^{\frac{1}{2}}$ is the sound speed at low frequencies. At present experimental results are sufficiently accurate to provide a meaningful comparison with the coefficient of ξ^2 in (4.85a), which we call the Burnett coefficient. The linear term in (4.85b) corresponds to the Kirchhoff absorption; its agreement with experiment is well known. The coefficient of ξ^3 in (4.85b), which we call the super-Burnett coefficient, is considerably more difficult to determine experimentally, so only a preliminary comparison is possible.

In figs. 6, 7, 8 the quantity $1 - v_0/v_{ph}$ is plotted as a function of ξ^2 for neon, argon and krypton. The points are the experimental results of Dr. Martin Greenspan, who estimates the accuracy to be about $\frac{1}{10}$ of a division on the vertical axis [14]. The straight lines labelled Burnett correspond to (4.85a) with Δ_1 evaluated at the indicated reduced temperatures for the Lennard-Jones model. The straight lines labelled Navier-Stokes correspond to (2.40a). The agreement between theory and experiment is satisfactory, including the curvature of the experimental points which, by comparing with the Maxwell model results (4.52), we see has the proper sign.

In fig. 9 the quantity $1 - 6\alpha v_0/7\,\omega\xi$ is plotted as a function of ξ^2 for neon. Again the points are the experimental results of Dr. Greenspan. The straight line labelled super-Burnett corresponds to (4.85b), the one labelled Navier-Stokes corresponds to (2.40b). The comparison between experiment and theory is at least suggestive of agreement, although the experimental accuracy is not so great as for the dispersion measurements.

This essential agreement between the results of perturbation theory applied to the linearized Boltzmann equation and experiment represents the best experimental confirmation of this equation of which we are aware. It would, of course, be extremely interesting to have experimental results sufficiently accurate to test the predicted reduced temperature dependence of the quantities Δ_1 and Δ_2.

8. REMARKS ON THE CONVERGENCE OF THE PERTURBATION EXPANSION

It has been pointed out by McLennan that a theorem due to Rellich on the perturbation of linear operators can be applied to prove the convergence of the E-series (4.19) in the case of hard spheres [15]. The Rellich theorem is as follows [16].

A self-adjoint linear operator $\boldsymbol{A}^{(0)}$ is perturbed by another self-adjoint linear operator $\boldsymbol{A}^{(1)}$, the resulting operator being

$$\boldsymbol{A} = \boldsymbol{A}^{(0)} + \epsilon\, \boldsymbol{A}^{(1)} . \tag{4.86}$$

We assume for arbitrary square integrable functions f these operators satisfy the condition:

$$\|A^{(1)} f\| \leqslant M [\|f\| + \|A^{(0)} f\|] , \tag{4.87}$$

where

$$\|f\| \equiv [(f, f)]^{\frac{1}{2}} , \tag{4.88}$$

and M is a fixed positive constant. Then, if $\lambda^{(0)}$ is an m-fold degenerate isolated eigenvalue of the unperturbed operator,

$$\boldsymbol{A}^{(0)} \psi_i^{(0)} = \lambda^{(0)} \psi_i^{(0)} , \qquad i = 1, 2, \ldots , m , \tag{4.89}$$

there will be m associated eigenvalues, counting multiplicities, of the operator $\boldsymbol{A}$,

$$\boldsymbol{A}\, \psi_i = \lambda_i(\epsilon)\, \psi_i , \qquad i = 1, 2, \ldots , m . \tag{4.90}$$

Moreover each $\lambda_i(\epsilon)$ is an analytic function of ϵ at $\epsilon = 0$ with $\lambda_i(0) = = \lambda^{(0)}$. Hence the perturbation series:

$$\lambda = \lambda^{(0)} + \epsilon \lambda^{(1)} + \epsilon^2 \lambda^{(2)} + \ldots \tag{4.91}$$

converges for ϵ sufficiently small.

Applying this theorem to the operators:

$$\boldsymbol{A}^{(0)} h \equiv \boldsymbol{J} h , \qquad \boldsymbol{A}^{(1)} h = c_z h , \tag{4.92}$$

we conclude that the E series (4.19) will converge for ϵ sufficiently small provided

$$\|c_z h\| < M [\|h\| + \|\boldsymbol{J} h\|] , \tag{4.93}$$

for arbitrary square integrable h.

As shown by McLennan, it is a simple matter to prove this condition is fulfilled for the hard sphere collision operator, for, since $c_z^2 < c^2$ one always has

$$\|c_z \varphi\| \leqslant \| c\, \varphi\| \quad . \tag{4.94}$$

But for hard spheres we can show that [17]

$$m(c) \geqslant \frac{\pi a^2}{\sigma} c \quad , \tag{4.95}$$

where $m(c)$ is the function given by (3.53). But, from (3.30),

$$m(c)h = \boldsymbol{K}h \ - \ \boldsymbol{J}h \ . \tag{4.96}$$

Hence, combining (4.94), (4.95) and (4.96) we get

$$\|c_z h\| \leqslant \frac{\sigma}{\pi a^2} \|\boldsymbol{K}h - \boldsymbol{J}h\| \ , \tag{4.97}$$

or, using the triangle inequality [18],

$$\|c_z h\| \leqslant \frac{\sigma}{\pi a^2} [\|\boldsymbol{K}h\| + \|\boldsymbol{J}h\|] \ . \tag{4.98}$$

Since the operator $\boldsymbol{K}$ is completely continuous for hard spheres (see chapter III, section 5), it follows that $\|\boldsymbol{K} h\|/\|h\|$ must be bounded and, hence, the condition (4.93) is fulfilled.

It is clear that this proof for hard spheres depends upon the Hil-Hilbert-Enskog decomposition (3.30) of $\boldsymbol{J}$, which is not possible for infinite range potentials such as the power law or Lennard-Jones models. However, the basic inequality (4.93) can certainly be discussed without decomposing $\boldsymbol{J}$.

For the case of Maxwell molecules, we can test the condition with Burnett functions ψ_{rl}. It will in fact be sufficient to consider only the case $l = 0$. From (4.9) we can see that

$$\| c_z \psi_{r0}\| = \left(\frac{4r+3}{6}\right)^{\frac{1}{2}} \ , \tag{4.99}$$

while for Maxwell molecules

$$\boldsymbol{J}\, \psi_{r0} = \lambda_{r0} \psi_{r0} \ , \tag{4.100}$$

where λ_{rl} is given by (3.95). Then, since the Burnett functions are normalized so $\|\psi_{r0}\| = 1$, the condition (4.93) becomes

$$\left(\frac{4r+3}{6}\right)^{\frac{1}{2}} \leqslant M(1 + |\lambda_{r0}|) \ . \tag{4.101}$$

But for large r we can show that [19]

$$\lambda_{r0} \sim -\frac{a^2}{\sigma}\left(\frac{6\pi\varphi_4}{kT}\right)^{\frac{1}{2}} \frac{\pi}{4}\, \Gamma(\tfrac{3}{4})\, r^{\frac{1}{4}} \ . \tag{4.102}$$

Hence, for sufficiently large r the right-hand side of (4.101) is smaller than the left-hand side for any fixed M and the Rellich condition is *not* satisfied. A similar argument holds for arbitrary repulsive power law forces.

We are therefore led to *suspect* that the perturbation expansion (4.19) does not converge for infinite range potentials. We must emphasize, however, that the Rellich theorem gives a sufficient but not necessary condition for convergence. Hence we really do not know the answer to the general convergence problem.

NOTES

1) C. S. Wang Chang and G. E. Uhlenbeck, *On the Propagation of Sound in Monoatomic Gases*, this *Study*. H. M. Mott-Smith, *A New Approach to the Kinetic Theory of Gases* (Lincoln Laboratory Group Report V-2, 1954).

2) To obtain this relation note first that since

$$Y_{l0}(\hat{c}) = \left(\frac{2l+1}{4\pi}\right)^{\frac{1}{2}} P_l(\cos\beta) , \tag{a}$$

where β is the polar angle of $\hat{c}$, then

$$\psi_{rl}(c) = \left[\frac{(2l+1)r!\,\Gamma(\frac{3}{2})}{\Gamma(r+l+\frac{3}{2})}\right]^{\frac{1}{2}} c^l\, L_r^{(l+\frac{1}{2})}(c^2) P_l(\cos\beta) .$$

Using the recursion relation for Legendre polynomials:

$$(2l+1)x P_l(x) = (l+1) P_{l+1}(x) + l\, P_{l-1}(x) , \tag{c}$$

[see A. R. Edmonds, *Angular Momentum in Quantum Mechanics* (Princeton University Press, Princeton, 1957) p. 23] we get

$$c_z\, \psi_{rl} = \left[\frac{r!\,\Gamma(\frac{3}{2})}{(2l+1)\Gamma(r+l+\frac{3}{2})}\right]^{\frac{1}{2}} \Big[(l+1)c^{l+1} L_r^{(l+\frac{1}{2})}(c^2) P_{l+1}(\cos\beta)$$

$$+ l\, c^{l+1} L_r^{(l+\frac{1}{2})}(c^2) P_{l-1}(\cos\beta)\Big] . \tag{d}$$

If now in the first term in the square brackets we use the recursion relation:

$$L_r^{(\alpha)}(z) = L_r^{(\alpha+1)}(z) - L_{r-1}^{(\alpha+1)}(z) , \tag{e}$$

and in the second term the recursion relation:

$$z\, L_r^{(\alpha)}(z) = (r+\alpha)\, L_r^{(\alpha-1)}(z) - (r+1)\, L_{r+1}^{(\alpha-1)}(z) \; , \tag{f}$$

and then use (b) we get the relation (4.9).

To obtain the relation (e) we start from (3.56) to write

$$L_r^{(\alpha+1)}(z) = \frac{1}{r!}\, z^{-\alpha-1}\, e^z \frac{d^r}{dz^r}\, e^{-z}\, z^{r+\alpha+1} \; . \tag{g}$$

If we note the identity:

$$\frac{d^r}{dz^r}\, z\, F(z) \equiv z\, \frac{d^r}{dz^r} F(z) + r\, \frac{d^{r-1}}{dz^{r-1}}\, F(z) \; , \tag{h}$$

for arbitrary $F(z)$, then from (g) we get

$$L_r^{(\alpha+1)}(z) = \frac{1}{r!}\, z^{-\alpha}\, e^z\, \frac{d^r}{dz^r}\, e^{-z}\, z^{r+\alpha} + \frac{1}{(r-1)!}\, z^{-\alpha+1}\, e^z\, \frac{d^{r-1}}{dz^{r-1}}\, e^{-z}\, z^{r+\alpha} , \tag{i}$$

which, referring to (3.56) is equivalent to (e). The recursion relation (f) is even simpler to obtain. We write

$$\begin{aligned} (r+1)\, L_{r+1}^{(\alpha-1)}(z) &= \frac{1}{r!}\, z^{-\alpha+1}\, e^z\, \frac{d^{r+1}}{dz^{r+1}}\, e^{-z}\, z^{r+\alpha} \\ &= \frac{1}{r!}\, z^{-\alpha+1}\, e^z\, \frac{d^r}{dz^r} \left[-e^{-z}\, z^{r+\alpha} + (r+\alpha)\, e^{-z}\, z^{r+\alpha-1} \right] , \end{aligned} \tag{j}$$

which, referring to (3.56) is equivalent to (f).

3) If the infinite determinant exists! See the remarks about the existence of this determinant at the end of section 2 of chapter V.

4) See S. Chapman and T. G. Cowling, *The Mathematical Theory of Non-Uniform Gases*, 2nd edition (Cambridge University Press, Cambridge, 1959) p. 162.

5) The most extensive calculations are those of Pekeris and his collaborators using a fast electronic computer. They obtain the dispersion relation from truncations as high as 485 × 485 for Maxwell molecules and as high as 150 × 105 for elastic spheres. See C. L. Pekeris, Z. Alterman, L. Finkelstein and K. Frankowski, Phys. Fluids 5 (1962) 1608.

6) This appears to be one of those "well-known facts" which appear rarely in print. See, however, H. A. Bethe and E. E. Salpeter, *Handbuch der Physik*, S. Flügge, ed., (Springer-Verlag, Berlin, 1957) Vol. XXXV, pp. 208-209.

7) These expressions are not as useful as they seem since they are strictly correct only for the *exact* $h^{(1)}_{\omega,k}$ and $h^{(2)}_{\omega,k}$. For approximate solutions of the first and second order equations the errors can be quite large.

8) See G. E. Uhlenbeck and G. W. Ford, *Lectures in Statistical Mechanics* (American Math. Soc., Providence, R. I., 1963) p. 107.

9) Chapman and Cowling, *op. cit.* See also D. Burnett, Proc. Lond. Math. Soc. 39 (1935) 385, where a proof of convergence is given for inverse power models.

10) For example, for the case of hard spheres, using the results of appendix A, we find

$$\frac{\left[a^{(1)}_{11}\right]_2}{\left[a^{(1)}_{11}\right]_1} = \frac{45}{44} = 1.02273 \ , \qquad \frac{\left[a^{(1)}_{11}\right]_3}{\left[a^{(1)}_{11}\right]_1} = \frac{60\,989}{59\,512} = 1.02481 \ ,$$

and, using the tables of L. Sirovich and J. K. Thurber, *Advances in Applied Mechanics*, Suppl. 3, Vol. 1, 1965, ed. J. H. Leeuw, we find

$$\frac{\left[a^{(1)}_{11}\right]_4}{\left[a^{(1)}_{11}\right]_1} = 1.02514 \ .$$

In the same way we find

$$\frac{\left[a^{(1)}_{02}\right]_4}{\left[a^{(1)}_{02}\right]_1} = 1.01600 \ .$$

Note that these are exactly the numerical coefficients in equation (2.44). See also Chapman and Cowling, *op. cit.*, chapter 10.

11) This is most easily demonstrated in the case of $E^{(3)}$, for which the identity (4.36) gives when we insert,

$$\left[h^{(1)}_{\omega,k}\right]_1 = \frac{1}{6^{\frac{1}{2}} J^1_{11}} \psi_{11} - \frac{1}{3^{\frac{1}{2}} J^2_{00}} \psi_{02} + \frac{1}{10^{\frac{1}{2}} J^1_{11}} \varphi_2 +$$

$$\left(\frac{1}{30^{\frac{1}{2}} J^2_{00}} - \frac{1}{120^{\frac{1}{2}} J^1_{11}}\right) \varphi_3 ,$$

the result

$$E^{(3)} = -\left(\tfrac{5}{6}\right)^{\frac{1}{2}}\left[\frac{1}{20(J^1_{11})^2} - \frac{1}{5J^1_{11}J^2_{00}} + \frac{4}{15(J^2_{00})^2}\right].$$

This is quite different from the expression (4.64) for $[E^{(3)}]_1$, and in fact results from that expression when we set the off diagonal matrix elements of $\boldsymbol{J}$ equal to zero.

12) See Chapman and Cowling, *op. cit.*, chapter 7. Also Uhlenbeck and Ford, *op. cit.*, chapter VI. In both cases the notation differs considerably from ours.

13) J. O. Hirschfelder, C. F. Curtis and R. B. Bird, *Molecular Theory of Gases and Liquids* (John Wiley and Sons, Inc., New York, 1964).

14) We wish to thank Dr. Greenspan for providing the tables of experimental results published in graphical form in *J. Acoust. Soc. Amer.* 28 (1956) 644.

15) J. A. McLennan, Phys. Fluids 8 (1965) 1580.

16) See F. Riesz and B. Sz.-Nagy, *Functional Analysis* (Frederick Ungar Publishing Co., New York, 1955) section 136. The Rellich theorem stated there is more general than the form we quote.

17) To see this note that

$$\frac{\mathrm{d}}{\mathrm{d}c}\frac{m(c)}{c} = -\frac{\pi^{\frac{1}{2}}a^2}{\sigma}\left[\mathrm{e}^{-c^2} + \frac{1}{c^2}\int_0^c \mathrm{d}u\,\mathrm{e}^{-u^2}\right] \leqslant 0 ,$$

which shows that $m(c)/c$ is a monotonic decreasing function of c, whose minimum value occurs at $c = \infty$. Hence, using (3.102) we get (4.95).

17) See Riesz and Sz.-Nagy, *op. cit.*, section 28.

18) To obtain this result we use the fact that for large r the integral in the expression (3.95) is dominated by the contribution from small angles where we may approximate:

$$f_4(\theta) \approx \frac{(3\pi)^{\frac{1}{2}}}{64}\sin^{-\frac{5}{2}}(\tfrac{1}{2}\theta) . \tag{a}$$

(See note 2, chapter III). We can therefore write for large r:

$$\lambda_{r0} \sim -\frac{\pi a^2}{16\sigma}\left(\frac{6\pi\,\varphi 4}{kT}\right)^{\frac{1}{2}}\int_0^{\pi}\mathrm{d}\theta\,\cos(\tfrac{1}{2}\theta)\,\sin^{-\frac{3}{2}}(\tfrac{1}{2}\theta)$$

$$\times\left[1 - \cos^{2r}(\tfrac{1}{2}\theta) - \sin^{2r}(\tfrac{1}{2}\theta)\right]. \tag{b}$$

But, putting $x = \sin^2(\frac{1}{2}\theta)$, the integral occurring here becomes

$$\int_0^1 dx x^{-\frac{5}{4}} \left[1-(1-x)^r - x^r\right] = 4r \int_0^1 dx x^{-\frac{1}{4}} \left[(1-x)^{r-1} - x^{r-1}\right], \qquad \text{(c)}$$

where we have integrated by parts. Recalling the Bernoulli integral (W. Magnus and F. Oberhettinger, *Formulas and Theorems for the Special Functions of Mathematical Physics* (Chelsea Publishing Co., New York, 1949) p. 4) this integral becomes

$$4r\left[\frac{\Gamma(\frac{3}{4})\,\Gamma(r)}{\Gamma(r+\frac{3}{4})} - \frac{1}{r-\frac{1}{4}}\right]. \qquad \text{(d)}$$

But for large r,

$$\frac{\Gamma(r)}{\Gamma(r+\frac{3}{4})} \sim r^{-\frac{3}{4}}. \qquad \text{(e)}$$

Hence the integral in (b) is for large r asymptotically equal to

$$4\,\Gamma(\tfrac{3}{4})\, r^{\frac{1}{4}}, \qquad \text{(f)}$$

which gives (4.102).

Chapter V

MODEL EQUATIONS

1. INTRODUCTION

As we have seen in chapter IV, the perturbation method can be used to calculate the terms in the expansion of the dispersion law in powers of the ratio of the mean free path to the wavelength; the number of terms we can calculate is limited, so to speak, only by our zeal. The question now arises of what happens when the wavelength becomes comparable to the mean free path so that this expansion is no longer useful. Here we should first remark that experiments, about which we shall have more to say in the next chapter, show large dispersion. In this chapter, however, we shall be concerned with the problem of how to obtain the dispersion law from the Boltzmann equation when the wave length and mean free path are comparable. What we shall do is to discuss certain model equations obtained by replacing the linearized collision operator by a much simplified operator. The resulting equations can be discussed completely and, we trust, give insight into the character of the solutions of the real equation.

There is an important question of principle in connection with this problem of obtaining sound dispersion from the Boltzmann equation. How do we distinguish the collective mode of oscillation we call sound from all the other normal modes of oscillation of the gas? At long wave lengths this is not a difficulty; the sound and heat conduction modes are those which grow out of the zero eigenvalues of the linearized collision operator. But at short wave lengths we can no longer treat the normal modes as perturbations of the static modes, and we must face the question. The answer is the chief aim of this chapter.

What we want to do is to investigate the character of *all* the normal modes of oscillation of the gas and to see what special character is possessed by the mode we call sound. Here there is a practical difficulty. The linearized collision operator has a complicated structure even for the simplest intermolecular force laws. However, it seems reasonable that the qualitative features of the normal modes depend only upon those properties of the collision operator $\boldsymbol{J}$ which are independent of the force law; i.e., that $\boldsymbol{J}$ has a five-fold degenerate zero eigenvalue and that all other eigenvalues are negative. We can, therefore, expect to learn much by replacing $\boldsymbol{J}$ by a simpler operator which has these properties and studying the normal modes of the resulting equation.

The general idea of what we do is as follows. The spectrum of the collision operator $\boldsymbol{J}$ is something like that shown in fig. 10,

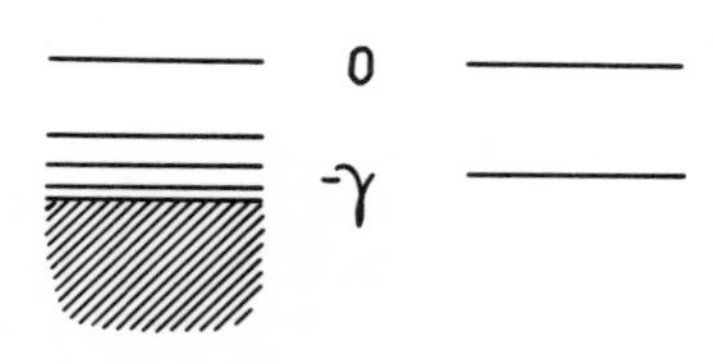

Fig. 10. The spectrum of the collision operator is indicated schematically on the left, that of the mutilated collision operator is shown on the right.

with a five-fold degenerate zero eigenvalue and a sequence of negative eigenvalues. In fact, as we saw in chapter III, the spectrum is unbounded from below and is perhaps in part continuous. We replace this collision operator with a *mutilated* collision operator $\boldsymbol{J}_1$ in which all the negative eigenvalues have been collapsed into a single eigenvalue of infinite degeneracy, as indicated in fig. 10. In fact, one gets:

$$\boldsymbol{J}_1 h = -\gamma h + \frac{\gamma}{\pi^{\frac{3}{2}}} \int \mathrm{d}\boldsymbol{c}_1 \, \mathrm{e}^{-c_1^2} \left[1 + 2\,\boldsymbol{c}\cdot\boldsymbol{c}_1 + \tfrac{2}{3}(c^2 - \tfrac{3}{2})(c_1^2 - \tfrac{3}{2})\right] h(\boldsymbol{c}_1) \quad , \tag{5.1}$$

where $-\gamma$ is the infinitely degenerate negative eigenvalue. One can readily verify that this mutilated collision operator has the desired properties: it annihilates any linear combination of the five conserved quantities $(1, \boldsymbol{c}, c^2)$ and multiplies any function which is orthogonal to these quantities by $-\gamma$. This mutilated collision operator can be considered as the first step in a successive approximation scheme for $\boldsymbol{J}$ in which successively more of the eigenvalues are treated exactly; as such it was proposed first by E. P. Gross and E. A. Jackson, and, independently, by M. Kac [1]).

Instead of discussing the linearized Boltzmann equation with the mutilated collision operator (5.1), we discuss first in the next section a simple one-dimensional analog of that equation. In the third section we turn to the more realistic three-dimensional problem, which although it is more complicated, requires arguments no different than those in the simpler one-dimensional problem. Finally, we make remarks on some of the extensive work which has been done with model equations.

2. NORMAL MODES OF A ONE-DIMENSIONAL MODEL EQUATION

Consider the equation:

$$\frac{\partial h}{\partial t} + c\,\frac{\partial h}{\partial x} = \boldsymbol{L}\,h \ , \tag{5.2}$$

where $h = h(x,c,t)$ and

$$\boldsymbol{L}h = -\,\gamma h(x,c,t) + \frac{\gamma}{\pi^{\frac{1}{2}}} \int_{-\infty}^{\infty} \mathrm{d}c_1\, \mathrm{e}^{-c_1^2}\,(1+2c\,c_1)\,h(x,c_1,t) \ . \tag{5.3}$$

The similarity of (5.2) to the linearized Boltzmann equation (3.14), with $\boldsymbol{L}$ the analog of the mutilated collision operator (5.1), is obvious [2]. Note that the linear operator $\boldsymbol{L}$ has a twofold degenerate zero eigenvalue, with eigenfunctions 1 and c, and an infinitely degenerate eigenvalue at $-\gamma$. That is, $\boldsymbol{L}$ annihilates any linear combination of 1 and c, while the result of operating with $\boldsymbol{L}$ on any function which is orthogonal to 1 and c is simply to multiply the function by $-\gamma$. Here we should remind ourselves that the scalar product in one dimension is

$$(f,g) = \pi^{-\frac{1}{2}} \int_{-\infty}^{\infty} \mathrm{d}c\ \mathrm{e}^{-c^2} f(c)\,g(c) \ . \tag{5.4}$$

We seek normal mode solutions of equation (5.2), i.e., we seek solutions of the form:

$$h\,(x,c,t) = h_{\omega,k}(c)\ \mathrm{e}^{i(kx-\omega t)} \ . \tag{5.5}$$

Inserting this assumed form for h in (5.2), using (5.3) and rearranging a little, we find

$$(\omega + i\gamma - kc)\,h_{\omega,k}(c) = i\gamma\,(\rho_{\omega,k} + 2c\ \varphi_{\omega,k}) \ , \tag{5.6}$$

where

$$\rho_{\omega,k} = (1, h_{\omega,k}) \; , \qquad \varphi_{\omega,k} = (c, h_{\omega,k}) \; , \tag{5.7}$$

are the one-dimensional analogs of the fluctuations in number of particles and mean flow velocity. In equation (5.6), since $\rho_{\omega,k}$ and $\varphi_{\omega,k}$ are independent of c, the dependence of the right-hand side upon c is explicit and we can therefore obtain an expression for $h_{\omega,k}$ by dividing through by the coefficient of $h_{\omega,k}$ on the left hand side. But here we must be careful! That coefficient can sometimes vanish, and we are not allowed to divide by zero. We must therefore distinguish two cases, depending upon whether the quantity

$$\bar{c} = (\omega + i\gamma)/k \tag{5.8}$$

is real or not [3].

Case 1. If $\bar{c}$ is real, then the solution of (5.6) is

$$h_{\omega,k} = A_{\omega,k}\, \delta(c - \bar{c}) - i\frac{\gamma}{k}(\rho_{\omega,k} + 2c\, \varphi_{\omega,k})\; \mathrm{P}\frac{1}{c - \bar{c}} \; , \tag{5.9}$$

where $A_{\omega,k}$ is an undetermined constant. Here $\delta(x)$ is the Dirac delta function and P stands for principal value. Thus the normal mode in this case is not a function, but rather a distribution. We get a trap for $\rho_{\omega,k}$ and $\varphi_{\omega,k}$ by inserting (5.9) in (5.7). We find

$$\rho_{\omega,k} = \pi^{-\frac{1}{2}} A_{\omega,k}\, \mathrm{e}^{-c^2} - i\frac{\gamma}{k}\pi^{-\frac{1}{2}}\, \mathrm{P}\int_{-\infty}^{\infty} \mathrm{d}c\, \mathrm{e}^{-c^2}\, \frac{\rho_{\omega,k} + 2c\, \varphi_{\omega,k}}{c - \bar{c}} \; ,$$

$$\varphi_{\omega,k} = \pi^{-\frac{1}{2}} A_{\omega,k}\, \bar{c}\, \mathrm{e}^{-\bar{c}^2} - i\frac{\gamma}{k}\pi^{-\frac{1}{2}}\, \mathrm{P}\int_{-\infty}^{\infty} \mathrm{d}c\, \mathrm{e}^{-c^2}\, \frac{c\,\rho_{\omega,k} + 2c^2\, \varphi_{\omega,k}}{c - \bar{c}} \; . \tag{5.10}$$

In appendix C, we introduce the function:

$$Y(x) = \pi^{-\frac{1}{2}}\, \mathrm{P}\int_{-\infty}^{\infty} \mathrm{d}t\, \mathrm{e}^{-t^2}\, \frac{1}{t - z} = -\,2\, \mathrm{e}^{-x^2} \int_0^x \mathrm{d}t\, \mathrm{e}^{t^2} \; . \tag{5.11}$$

Using this definition, we can write equations (5.10) in the form:

$$\left[1 + i\frac{\gamma}{k} Y(\bar{c})\right] \rho_{\omega,k} + 2i\frac{\gamma}{k}\left[1 + \bar{c}\, Y(\bar{c})\right]\, \varphi_{\omega,k} = \pi^{-\frac{1}{2}} A_{\omega,k}\, \mathrm{e}^{-\bar{c}^2} \; ,$$

$$i\frac{\gamma}{k}\,[1+\bar{c}Y(\bar{c})]\;\rho_{\omega,k}+\left\{1+2i\frac{\gamma}{k}\bar{c}\,[1+\bar{c}Y(\bar{c})]\right\}\;\varphi_{\omega,k}=\pi^{-\frac{1}{2}}A_{\omega,k}\,\bar{c}\,\mathrm{e}^{-\bar{c}^2}. \tag{5.12}$$

This is a pair of inhomogeneous linear equations to be solved for the two amplitudes $\rho_{\omega,k}$ and $\varphi_{\omega,k}$. The solution is

$$\rho_{\omega,k}=\frac{\pi^{-\frac{1}{2}}\,\mathrm{e}^{-\bar{c}^2}A_{\omega,k}}{1+i\frac{\gamma}{k}[2\bar{c}+(2\bar{c}^2+1)Y(\bar{c})]+2\frac{\gamma^2}{k^2}[1+\bar{c}Y(\bar{c})]},$$

$$\varphi_{\omega,k}=\frac{\pi^{-\frac{1}{2}}\,\mathrm{e}^{-\bar{c}^2}(\bar{c}-i\frac{\gamma}{k})A_{\omega,k}}{1+i\frac{\gamma}{k}[2\bar{c}+(2\bar{c}^2+1)Y(\bar{c})]+2\frac{\gamma^2}{k^2}[1+\bar{c}Y(\bar{c})]}\,. \tag{5.13}$$

Inserting this solution in (5.9) we obtain an explicit solution of the normal mode problem for this case. Note there is no functional relation between ω and k as a condition for the existence of these normal modes; i.e., there is no dispersion law. If we think of k being given and real, then there is a continuum of normal mode solutions, all with imaginary part $-\gamma$. These modes, therefore, are all damped in time with a characteristic relaxation time γ^{-1}.

Case 2: If $\bar{c}$ is not real, then we can simply divide both sides of (5.6) by the coefficient of $h_{\omega,k}$, which is non-zero, and obtain:

$$h_{\omega,k}=-i\,\frac{\gamma}{k}\,\frac{\rho_{\omega,k}+2c\,\varphi_{\omega,k}}{c-\bar{c}}\,. \tag{5.14}$$

Again, we determine $\rho_{\omega,k}$ and $\varphi_{\omega,k}$ by inserting this expression in (5.7). The result is:

$$\rho_{\omega,k}=-i\,\frac{\gamma}{k}\,\pi^{-\frac{1}{2}}\int_{-\infty}^{\infty}\mathrm{d}c\;\mathrm{e}^{-c^2}\,\frac{\rho_{\omega,k}+2c\,\varphi_{\omega,k}}{c-\bar{c}}\,,$$

$$\varphi_{\omega,k}=-i\,\frac{\gamma}{k}\,\pi^{-\frac{1}{2}}\int_{-\infty}^{\infty}\mathrm{d}c\;\mathrm{e}^{-c^2}\,\frac{c\,\rho_{\omega,k}+2c^2\,\varphi_{\omega,k}}{c-\bar{c}}\,, \tag{5.15}$$

which is a pair of homogeneous equations for the determination of $\rho_{\omega,k}$ and $\varphi_{\omega,k}$. If we introduce the function

$$Z(z)=\pi^{-\frac{1}{2}}\int_{-\infty}^{\infty}\mathrm{d}c\;\mathrm{e}^{-c^2}\,\frac{1}{c-z}\,, \tag{5.16}$$

which is discussed in appendix C, we can write these equations in the form:

$$\left[1+i\frac{\gamma}{k}Z(\bar{c})\right]\rho_{\omega,k}+2i\frac{\gamma}{k}\left[1+\bar{c}Z(\bar{c})\right]\varphi_{\omega,k}=0\ ,$$

$$i\frac{\gamma}{k}\left[1+\bar{c}Z(\bar{c})\right]\rho_{\omega,k}+\left\{1+2i\frac{\gamma}{k}\bar{c}\left[1+\bar{c}Z(\bar{c})\right]\right\}\varphi_{\omega,k}=0\ . \tag{5.17}$$

A non-trivial solution of these homogeneous equations exists if and only if the determinant of the coefficients vanishes, i.e.,

$$F(\omega,k)\equiv 1+i\frac{\gamma}{k}\left[2\bar{c}+(2\bar{c}^2+1)Z(\bar{c})\right]+2\frac{\gamma^2}{k^2}\left[1+\bar{c}Z(c)\right]=0\ . \tag{5.18}$$

This, as we see, is a dispersion law; for any value of ω and k satisfying this relation we can construct a non-trivial solution of (5.17) which, when inserted in (5.14) gives an explicit solution of the normal mode problem. Note that these solutions are well behaved *functions*. Also, since the function $Z(\bar{c})$ is analytic everywhere except on the real axis, there can be, for given k, only a discrete set of values of ω fulfilling this dispersion relation, corresponding to a discrete set of normal modes. Among these normal modes is the sound-like mode we seek.

We must study now the zeros of the dispersion relation (5.18). We do this by considering γ/k real and then determining the complex values of $\bar{c}$ for which F vanishes. To see the general picture, we use the *winding theorem*: the number of zeros of F in a region of the complex $\bar{c}$-plane in which F is an analytic function of $\bar{c}$ is equal to the number of times the representative point F in the complex F-plane encircles the origin as $\bar{c}$ is carried around the boundary of the region in the $\bar{c}$-plane [4]. Consider first the application of this theorem to the upper half $\bar{c}$-plane. The boundary of this region consists of the real axis together with a semi-circle at infinity. But for $|\bar{c}|$ large $Z(\bar{c})$ has the asymptotic expansion:

$$Z(\bar{c})\sim-\sum_{n=0}^{\infty}\frac{\Gamma(n+\frac{1}{2})}{\Gamma(\frac{1}{2})}\,\bar{c}^{\,-2n-1}\ , \tag{5.19}$$

from which we can see that the point F is stationary at $F=1$ as $\bar{c}$ is carried along the large semi-circle. Along the real axis, we use the fact that, for real x,

$$\lim_{\epsilon \to 0^+} Z(x+i\epsilon) = Y(x) + i\,\pi^{\frac{1}{2}}\,\mathrm{e}^{-x^2}\,, \tag{5.20}$$

where $Y(x)$ is given by (5.11). Hence, as $\bar{c}$ is carried from $-\infty$ to ∞ along the real axis, the real and imaginary parts of F are given by

$$\mathrm{Re}\,F = 1 - \frac{\gamma}{k}\pi^{\frac{1}{2}}(2\bar{c}^2+1)\mathrm{e}^{-\bar{c}^2} + 2\frac{\gamma^2}{k^2}[1+\bar{c}\,Y(\bar{c})]\,,$$

$$\mathrm{Im}\,F = \frac{\gamma}{k}[2\bar{c}+(2\bar{c}^2+1)Y(\bar{c})] + 2\frac{\gamma^2}{k^2}\pi^{\frac{1}{2}}\bar{c}\,\mathrm{e}^{-\bar{c}^2}\,. \tag{5.21}$$

It remains only to plot $\mathrm{Im}\,F$ versus $\mathrm{Re}\,F$, using tables of the function Y, and count the number of times the resulting closed curve encloses the origin. What we find is that there is a critical value k_0 of the wave number k, numerically $k_0/\gamma = 1.95259...$. For k less than k_0, i.e., at long wavelengths, there are exactly two zeros; a typical F-trajectory for k in this range is shown in fig.11. On the other hand, for k greater than k_0 there are no zeros; a typical F-trajectory for this case is shown in fig. 12.

When we apply the winding theorem to the lower half plane we must realize that the function $Z(\bar{c})$ in the lower half $\bar{c}$-plane is not obtained by analytic continuation from the upper half-plane. In par-

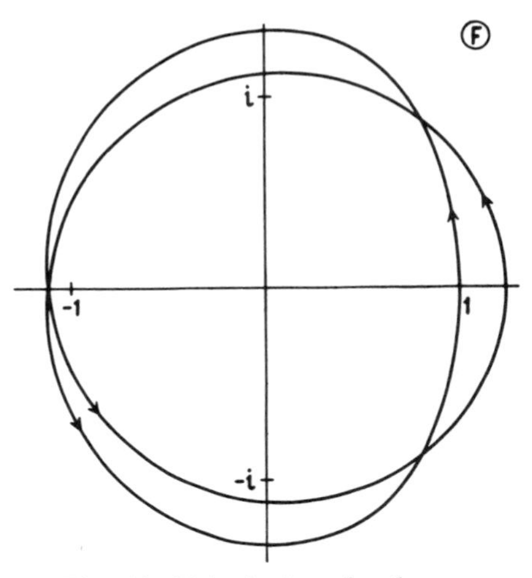

Fig. 11. F-trajectory $k < k_0$.

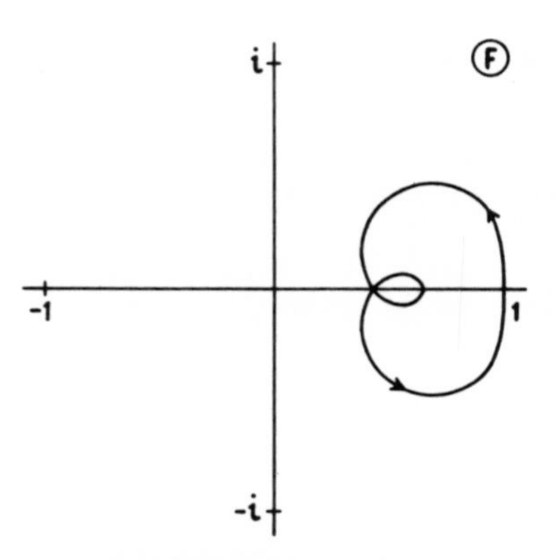

Fig. 12. F-trajectory $k > k_o$.

ticular, for real x,

$$\lim_{\epsilon \to 0^+} Z(x - i\epsilon) = Y(x) - i\,\pi^{\frac{1}{2}}\, e^{-x^2} \quad . \tag{5.22}$$

Hence, as $\bar{c}$ is carried from ∞ to $-\infty$ along the real axis, the real and imaginary parts of F are given not by (5.21), but rather by

$$\operatorname{Re} F = 1 + \frac{\gamma}{k}\pi^{\frac{1}{2}}\,(2\bar{c}^2 + 1)\; e^{-\bar{c}^2} + 2\,\frac{\gamma^2}{k^2}\,[1 + \bar{c}\,Y(\bar{c})] \;,$$

$$\operatorname{Im} F = \frac{\gamma}{k}\,[2\bar{c} + (2\bar{c}^2 + 1)\; Y(\bar{c})] - 2\,\frac{\gamma^2}{k^2}\,\pi^{\frac{1}{2}}\;\bar{c}\; e^{-\bar{c}^2} \quad . \tag{5.23}$$

In this case the F-trajectory is always of the form shown in fig.13,

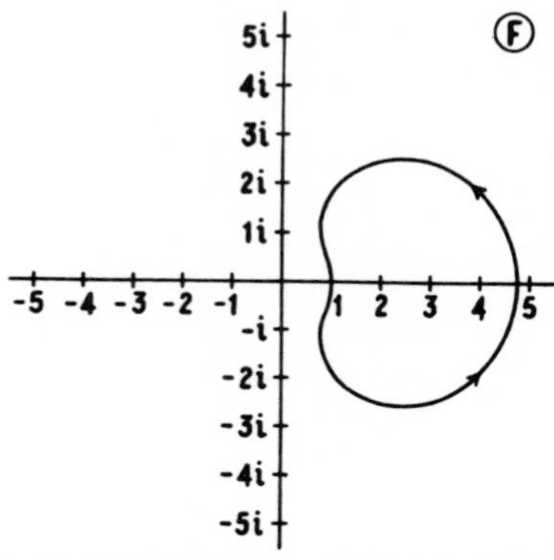

Fig. 13. F-trajectory for the lower half-plane.

which, since the curve does not enclose the origin, indicates that there are never any zeros of F in the lower half plane. We conclude therefore, that the dispersion law (5.18) has zeros only if k is less than k_0, in which case there are exactly two, lying in the upper half $\bar{c}$-plane.

We must now investigate more carefully the path of these zeros as a function of k. Consider first the case of small k, i.e., long wavelengths. Here $\bar{c}$, given by (5.8), is large in absolute value and we can use the asymptotic expansion (5.19) for Z. Inserting this expansion in (5.18) we can without too much effort find the first few terms in the expansions of the zeros for small k. The results are

$$\bar{c} = i\left(\frac{\gamma}{k} - \tfrac{1}{2}\frac{k}{\gamma} + \tfrac{1}{2}\frac{k^3}{\gamma^3} - 2\frac{k^5}{\gamma^5} + \frac{51}{4}\frac{k^7}{\gamma^7} + \dots\right)$$

$$\pm 2^{-\frac{1}{2}}\left(1 + \frac{1}{4}\frac{k^2}{\gamma^2} - \frac{33}{32}\frac{k^4}{\gamma^4} + \frac{801}{128}\frac{k^6}{\gamma^6} - \frac{99\,013}{2048}\frac{k^8}{\gamma^8} + \dots\right) . \tag{5.24}$$

Note that the two zeros have the same imaginary part and the real parts have opposite signs; it is not difficult to see that this must be so directly from the dispersion law. For intermediate values of k we determine the zeros numerically, using tables [5)] of the function Z. We find that as k increases the zeros move rapidly toward the real axis along the path shown in fig. 14. The zeros reach the axis when $k = k_0$, at which point they disappear.

The normal modes associated with the pair of zeros of the dispersion law correspond to sound-like propagating modes. We obtain the following expansion of ω in powers of k for these modes from (5.24) and the definition (5.8) of $\bar{c}$:

$$\frac{\omega}{\gamma} = \pm 2^{-\frac{1}{2}}\left(\frac{k}{\gamma} + \frac{1}{4}\frac{k^3}{\gamma^3} - \frac{33}{32}\frac{k^5}{\gamma^5} + \frac{801}{128}\frac{k^7}{\gamma^7} - \frac{99\,103}{2048}\frac{k^9}{\gamma^9} + \dots\right)$$

$$-i\left(\frac{1}{2}\frac{k^2}{\gamma^2} - \frac{1}{2}\frac{k^4}{\gamma^4} + 2\frac{k^6}{\gamma^6} - \frac{51}{4}\frac{k^8}{\gamma^8} + \dots\right) . \tag{5.25}$$

The similarity of this expansion with the expansions of the dispersion law for sound in three dimensions is obvious. From the numerical data on the motion of the zeros of F, we can determine the phase velocity

$$v_{\mathrm{ph}} = \mathrm{Re}\left\{\frac{\omega}{k}\right\} \tag{5.26}$$

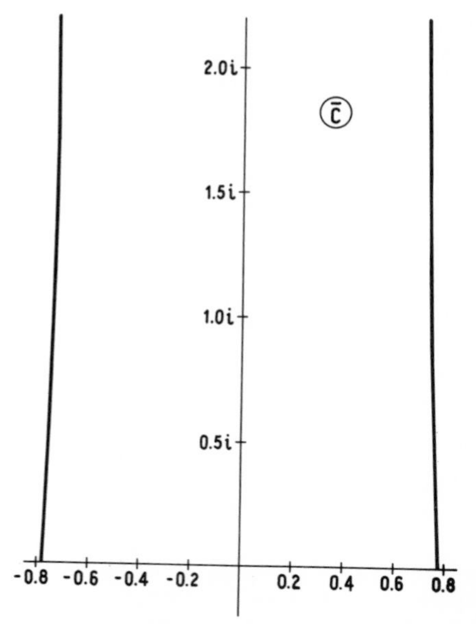

Fig. 14. Path of the zeros in the $\bar{c}$-plane for the one-dimensional model.

and the relaxation time

$$1/\tau = - \operatorname{Im} \{\omega\} \tag{5.27}$$

as functions of k for all $k < k_0$. These are shown in fig. 15 and fig. 16.

At this point we make a number of remarks:

1) The normal modes for this model equation fall into two classes. The first consists of the normal modes associated with the continuum of values of ω whose imaginary part is $-\gamma$. For these normal modes there is no dispersion law relating ω and k. The second class consists of the normal modes associated with the discrete zeros of the dispersion law (5.18). Here there is a dispersion law since these zeros move along a continuous curve as k is varied. The sound-like modes belong to this second class. For our model equation this answers the question posed in the introduction of how to distinguish the sound modes from the infinity of possible normal modes of the gas; the sound modes are those discrete modes which grow out of the zero eigenvalues of the collision operator.

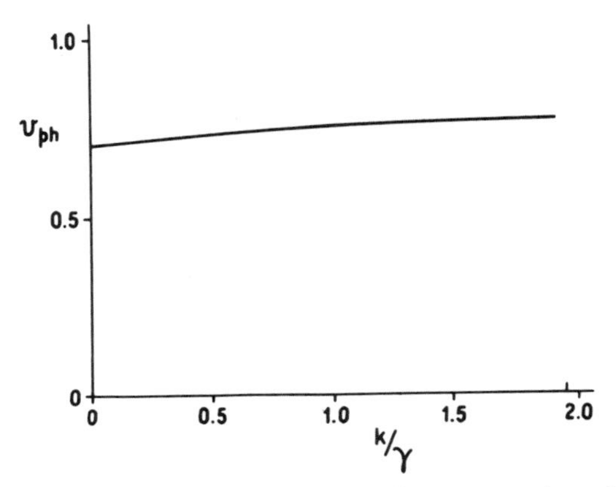

Fig. 15. The phase velocity as a function of wave number for the one-dimensional model.

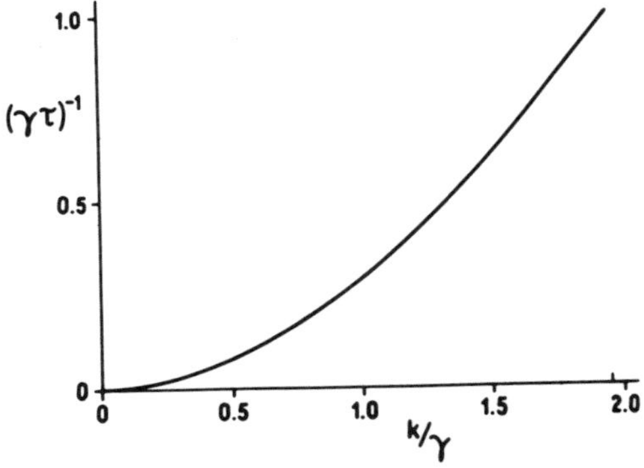

Fig. 16. The relaxation time as a function of wave number for the one-dimensional model.

2) Some authors call the modes of class 1, those associated with the continuum, single-particle modes. We would propose instead to call them Knudsen modes because of their close relation with the normal modes in the Knudsen regime, where the mean free path is long compared with the wave length. In this regime the effects of collisions may be neglected and the motion is governed by the free streaming of the gas. For our model this corresponds to setting L equal to zero. The corresponding normal modes are

$$h_{\omega,k} = A_{\omega,k}\,\delta(c - \bar{c}) \ , \tag{5.28}$$

where Im $\omega = -\gamma$ and $\bar{c}$ is given by (5.8). The relation to the normal modes of class 1 is obvious.

3) The sudden disappearance of the discrete modes at a critical value of k is probably a general phenomenon. If we modify our one dimensional model equation by replacing $\boldsymbol{L}$ by an operator which has, in addition to the doubly degenerate zero eigenvalue and the infinitely degenerate eigenvalue at $-\gamma$, one or more eigenvalues of finite multiplicity in the gap between 0 and $-\gamma$, then we find that for each added eigenfunction a new discrete normal mode appears. These new modes are all more highly damped than the sound mode, indeed they have a finite relaxation time at infinite wave lengths, but like the sound modes they each disappear at a characteristic critical frequency. At sufficiently short wave lengths only the Knudsen modes remain.

We turn as a last item in this section to a discussion of the method of Chang and Uhlenbeck as applied to this model. We recall that with their method one first expands $h_{\omega,k}$ in terms of a complete orthonormal set of functions of the velocity, in their case the Burnett functions, in our case the one-dimensional analog of the Burnett functions:

$$\psi_n(c) \equiv (2^n\, n!)^{-\frac{1}{2}} H_n(c) \ , \tag{5.29}$$

where $H_n(c)$ is the Hermite polynomial (see appendix C). Hence, we write

$$h_{\omega,k}(c) = \sum_{n=0}^{\infty} a_n \psi_n(c) \ . \tag{5.30}$$

Inserting this expansion in the equation (5.6) for the normal modes, and using the recursion relation:

$$c\psi_n = (\tfrac{1}{2}(n+1))^{\frac{1}{2}}\, \psi_{n+1} + (\tfrac{1}{2}n)^{\frac{1}{2}}\, \psi_n, \tag{5.31}$$

we get the following infinite set of coupled linear equations for the a_n:

$$\left[\bar{c} - i\frac{\gamma}{k}(\delta_{n,0} + \delta_{n,1})\right] a_n - (\tfrac{1}{2}n)^{\frac{1}{2}} a_{n-1} - (\tfrac{1}{2}(n+1))^{\frac{1}{2}} a_{n+1} = 0 \ . \tag{5.32}$$

Here we have introduced $\bar{c}$ defined by (5.8). The Chang determinant is the infinite determinant of the coefficients of these equations, which is set equal to zero as a condition of existence of solutions. It has the form [6]:

$$D \equiv \begin{vmatrix} \bar{c}-i\gamma/k & -(\frac{1}{2})^{\frac{1}{2}} & 0 & 0 & 0 & \cdots \\ -(\frac{1}{2})^{\frac{1}{2}} & \bar{c}-i\gamma/k & -(\frac{2}{2})^{\frac{1}{2}} & 0 & 0 & \cdots \\ 0 & -(\frac{2}{2})^{\frac{1}{2}} & \bar{c} & -(\frac{3}{2})^{\frac{1}{2}} & 0 & \cdots \\ 0 & 0 & -(\frac{3}{2})^{\frac{1}{2}} & \bar{c} & -(\frac{4}{2})^{\frac{1}{2}} & \cdots \\ 0 & 0 & 0 & -(\frac{4}{2})^{\frac{1}{2}} & \bar{c} & \cdots \\ \cdots & \cdots & \cdots & \cdots & \cdots & \\ \cdots & \cdots & \cdots & \cdots & \cdots & \end{vmatrix} \tag{5.33}$$

According to the method of Chang and Uhlenbeck, this determinant is successively approximated by truncation: the n'th approximation, D_n, is the $n+1$ by $n+1$ principal minor in the upper left hand corner. Thus

$$D_1 = (\bar{c}-i\gamma/k)^2 - \tfrac{1}{2} ,$$

$$D_2 = \bar{c}\ (\bar{c}-i\gamma/k)^2 - \tfrac{1}{2}\ \bar{c} - \tfrac{3}{2}\ (\bar{c}-i\gamma/k) , \tag{5.34}$$

and so on. Setting D_n equal to zero gives us the n'th approximation to the dispersion relation, e.g. in the first approximation we get

$$\bar{c} = - i\ \gamma/k \pm 2^{-\frac{1}{2}} , \tag{5.35}$$

which correctly gives the first two terms in the expansion (5.24).

The relation of these successive approximations of the Chang determinant to the exact dispersion law (5.18) is shown by the identity:

$$D_{n-1} = 2^{-n} H_n(\bar{c})\ F_n , \tag{5.36}$$

where H_n is the Hermite polynomial and F_n is obtained from (5.18) by replacing the function $Z(c)$ by

$$Z_n(\bar{c}) = - 2\ \frac{G_n(\bar{c})}{H_n(\bar{c})} . \tag{5.37}$$

Here $G_n(\bar{c})$ is a polynomial related to the Hermite polynomial and discussed in appendix C. To prove the identity (5.36) we note that by expanding D_n in terms of the minors of the last row we obtain the recursion relation:

$$D_n = \bar{c}\, D_{n-1} - \tfrac{1}{2} n\, D_{n-2} \ . \tag{5.38}$$

On the other hand we can show from the recursion relations (C.14) for H_n and G_n that the right hand side of (5.36) satisfies this same recursion relation. Finally, it is a simple matter to verify this identity for $n = 1$ and $n = 2$, from which the general result follows.

The significance of the identity (5.36) stems from the fact that

$$\lim_{n \to \infty} Z_n(\bar{c}) = Z(\bar{c}) \ , \tag{5.39}$$

for $\operatorname{Im} \bar{c} \neq 0$. It follows that

$$\lim_{n \to \infty} F_n = F \ , \tag{5.40}$$

where F is given by (5.18). Since for large n we have [7]

$$H_n(\bar{c}) \sim 2^{\frac{1}{2}(n+1)} (n/\mathrm{e})^{\frac{1}{2}n}\, \mathrm{e}^{\frac{1}{2}\bar{c}^2} \cos[(2n+1)^{\frac{1}{2}} \bar{c} - \tfrac{1}{2} n\pi] \ , \tag{5.41}$$

we can conclude that the limit for large n of D_n does not exist. That is, the successive truncations of the Chang determinant do not converge, which means that the Chang determinant itself does not exist. On the other hand, the Hermite polynomial has only real zeros, from which we conclude that the non-real zeros of D_n are the non-real zeros of F_n, which converge to the non-real zeros of F discussed above. These zeros of F correspond to the dispersion law for the sound-like modes, so we can conclude that the successive approximations to the dispersion law for the sound modes obtained by setting the successive truncations of the Chang determinant equal to zero converge to the exact dispersion law. Hence, in this sense at least, the method of Chang and Uhlenbeck leads to the correct dispersion law for the sound-like modes. On the other hand, it tells us nothing about the Knudsen modes. It seems not unlikely that the same conclusions will be valid for the method of Chang and Uhlenbeck as applied to a real gas.

3. THREE-DIMENSIONAL MODEL EQUATIONS

We consider here the dispersion of sound using the mutilated collision operator. Since the discussion is essentially the same as that for the one-dimensional model discussed in section 2, only more complicated, we give only a brief sketch. We replace the collision operator in the equation (4.3) for the normal modes of the linearized Boltzmann equation with the mutilated collision operator (5.1) to get the equation:

$$\left[-i\left(\frac{m}{2kT}\right)^{\frac{1}{2}} \omega + i\,\boldsymbol{k}\cdot\boldsymbol{c}\right] h_{\omega,\boldsymbol{k}}(\boldsymbol{c}) = -n\sigma\gamma h_{\omega,\boldsymbol{k}}(\boldsymbol{c}) + n\sigma\gamma\left[\rho_{\omega,\boldsymbol{k}} + 2\,\boldsymbol{c}\cdot\boldsymbol{\varphi}_{\omega,\boldsymbol{k}} + \tfrac{2}{3}(c^2 - \tfrac{3}{2})\left(2\,\mathcal{E}_{\omega,\boldsymbol{k}} - \tfrac{3}{2}\rho_{\omega,\boldsymbol{k}}\right)\right], \tag{5.42}$$

where

$$\rho_{\omega,\boldsymbol{k}} \equiv (1, h_{\omega,\boldsymbol{k}}) \equiv \pi^{-\frac{3}{2}} \int \mathrm{d}\boldsymbol{c}\; \mathrm{e}^{-c^2} h_{\omega,\boldsymbol{k}}(\boldsymbol{c}) \;, \tag{5.43a}$$

$$\boldsymbol{\varphi}_{\omega,\boldsymbol{k}} \equiv (\boldsymbol{c}, h_{\omega,\boldsymbol{k}}) \equiv \pi^{-\frac{3}{2}} \int \mathrm{d}\boldsymbol{c}\; \mathrm{e}^{-c^2} \boldsymbol{c}\, h_{\omega,\boldsymbol{k}}(\boldsymbol{c}) \;, \tag{5.43b}$$

$$\mathcal{E}_{\omega,\boldsymbol{k}} \equiv (\tfrac{1}{2}c^2, h_{\omega,\boldsymbol{k}}) \equiv \pi^{-\frac{3}{2}} \int \mathrm{d}\boldsymbol{c}\; \mathrm{e}^{-c^2}\, \tfrac{1}{2}c^2\, h_{\omega,\boldsymbol{k}}(\boldsymbol{c}) \;, \tag{5.43c}$$

which we call the fluctuation amplitudes, correspond to fluctuations in number density, local flow velocity, and energy in the gas. We introduce the dimensionless quantities:

$$E = -i\left(\frac{m}{2kT}\right)^{\frac{1}{2}} \frac{\omega}{n\sigma} \,, \qquad \epsilon = -i\,\frac{k}{n\sigma} \,, \tag{5.44}$$

corresponding to the quantities introduced in (4.5). If we put

$$\bar{c} \equiv (E + \gamma)/\epsilon \;, \tag{5.45}$$

we can write (5.42) in the form:

$$(\bar{c} - \hat{k}\cdot c)\, h_{\omega,k} = \frac{\gamma}{\epsilon}\Big[\rho_{\omega,k} + 2\, c\cdot\varphi_{\omega,k} + \tfrac{2}{3}(c^2 - \tfrac{3}{2})(2\mathcal{E}_{\omega,k} - \tfrac{3}{2}\rho_{\omega,k})\Big]. \tag{5.46}$$

Here the right hand side is explicit in its dependence upon c, so we need only divide by the factor multiplying $h_{\omega,k}$. However, if $\bar{c}$ is real this factor will vanish for certain values of c, so we again must distinguish two cases.

Case 1. If $\bar{c}$ is real, then the solution of (5.46) is

$$h_{\omega,k}(c) = A_{\omega,k}\,\delta(\bar{c} - \hat{k}\cdot c) - \frac{\gamma}{\epsilon}\,\mathrm{P}\,\frac{\rho_{\omega,k} + 2\, c\cdot\varphi_{\omega,k} + \frac{2}{3}(c^2 - \frac{3}{2})(2\mathcal{E}_{\omega,k} - \frac{3}{2}\rho_{\omega,k})}{\hat{k}\cdot c - \bar{c}} \tag{5.47}$$

where $A_{\omega,k}$ is an arbitrary complex amplitude. The five fluctuation amplitudes are determined by inserting (5.47) in (5.43). The result is a set of five *inhomogeneous* linear equations, which may in general be solved to express $\rho_{\omega,k}$, $\varphi_{\omega,k}$ and $\mathcal{E}_{\omega,k}$ in terms of $A_{\omega,k}$. Thus we can produce a complete and explict solution for the normal modes in this case. These modes, which, again, we term Knudsen modes, are *distributions*, not functions. They correspond to a *continuum* of values of ω for each value of k, i.e. all ω with $\mathrm{Im}\,\omega = -n\sigma\gamma(2kT/m)^{\frac{1}{2}}$. Hence there is no disperion law relating ω and k for these modes.

Case 2. If $\bar{c}$ is not real, then the solution of (5.46) is

$$h_{\omega,k}(c) = -\frac{\gamma}{\epsilon}\,\frac{\rho_{\omega,k} + 2c\cdot\varphi_{\omega,k} + \frac{2}{3}(c^2 - \frac{3}{2})(2\mathcal{E}_{\omega,k} - \frac{3}{2}\rho_{\omega,k})}{\hat{k}\cdot c - \bar{c}}. \tag{5.48}$$

Here again we determine the five fluctuation amplitudes by inserting (5.48) in (5.43). The result [8] is the following set of five coupled *homogeneous* linear equations for the determination of $\rho_{\omega,k}$, $\varphi_{\omega,k}$, $\mathcal{E}_{\omega,k}$:

$$\{\tfrac{\epsilon}{\gamma} - [\bar{c} + (\bar{c}^2 - \tfrac{3}{2})\, Z(\bar{c})]\}\,\rho_{\omega,k} + 2\,[1 + \bar{c}Z(\bar{c})]\,\hat{k}\cdot\varphi_{\omega,k} + \tfrac{4}{3}\,[\bar{c} + (\bar{c}^2 - \tfrac{1}{2})\, Z(\bar{c})]\,\mathcal{E}_{\omega,k} = 0\,, \tag{5.49a}$$

$$- [\bar{c}^2 - 1 + \bar{c}(\bar{c}^2 - \tfrac{3}{2}) Z(\bar{c})] \rho_{\omega,\boldsymbol{k}} + \{\tfrac{\epsilon}{\gamma} + 2\bar{c}[1 + \bar{c}Z(\bar{c})]\} \hat{\boldsymbol{k}}\cdot\boldsymbol{\varphi}_{\omega,\boldsymbol{k}}$$

$$+ \tfrac{4}{3}\bar{c}[\bar{c} + (\bar{c}^2 - \tfrac{1}{2}) Z(\bar{c})] \mathcal{E}_{\omega,\boldsymbol{k}} = 0 \;, \tag{5.49b}$$

$$- \tfrac{1}{2}[\bar{c}^3 + (\bar{c}^4 - \tfrac{1}{2}\bar{c}^2 - \tfrac{1}{2}) Z(\bar{c})] \rho_{\omega,\boldsymbol{k}} + [\bar{c}^2 + \tfrac{3}{2} + \bar{c}(\bar{c}^2 + 1) Z(\bar{c})] \hat{\boldsymbol{k}}\cdot\boldsymbol{\varphi}_{\omega,\boldsymbol{k}}$$

$$+ \{\tfrac{\epsilon}{\gamma} + \tfrac{2}{3}[\bar{c}(\bar{c}^2 + 1) + (\bar{c}^4 + \tfrac{1}{2}\bar{c}^2 + \tfrac{1}{2}) Z(\bar{c})]\} \mathcal{E}_{\omega,\boldsymbol{k}} = 0 \;, \tag{5.49c}$$

$$[\tfrac{\epsilon}{\gamma} + Z(\bar{c})] (\boldsymbol{\varphi}_{\omega,\boldsymbol{k}} - \hat{\boldsymbol{k}}\cdot\boldsymbol{\varphi}_{\omega,\boldsymbol{k}} \hat{\boldsymbol{k}}) = 0 \;. \tag{5.49d}$$

Consider first the last equation (5.49d) which is an equation for the amplitude of the transverse velocity fluctuations. There is a non-trivial solution if and only if

$$\frac{\epsilon}{\gamma} + Z(\bar{c}) = 0 \;, \tag{5.50}$$

which is the dispersion law for the transverse sound modes. For small ϵ, i.e. long wavelengths, the solution may be obtained as an expansion in powers of ϵ by using the asymptotic expansion (5.19). We express the answer in terms of the dimensionless variable E, and find

$$E = \frac{\epsilon^2}{2\gamma} + \frac{\epsilon^4}{4\gamma^3} + \frac{7\epsilon^6}{8\gamma^5} + \ldots \;. \tag{5.51}$$

One can show from (5.50) that for increasing k as ϵ moves down the imaginary axis the point E moves along the negative real axis until k reaches the critical value $k_0 = \pi^{\frac{1}{2}} n\sigma\gamma$, at which point this transverse sound mode disappears. Since these transverse sound modes are of relatively little interest, we say no more about them.

Next consider the three equations (5.49 a-c), which form three coupled linear homogeneous equations for the determination of the fluctuation amplitudes of the density, $\rho_{\omega,\boldsymbol{k}}$, longitudinal flow velocity, $\hat{\boldsymbol{k}}\cdot\boldsymbol{\varphi}_{\omega,\boldsymbol{k}}$, and energy density, $\mathcal{E}_{\omega,\boldsymbol{k}}$. There is a non-trivial solution of these equations if and only if the determinant of the coefficients vanishes, i.e.,

$$F(\omega, k) \equiv \left(\frac{\epsilon}{\gamma}\right)^3 + \left(\frac{\epsilon}{\gamma}\right)^2 \Phi_1(\bar{c}) + \frac{\epsilon}{\gamma}\Phi_2(\bar{c}) + \Phi_3(\bar{c}) = 0 \;, \tag{5.52}$$

where we have introduced

$$\Phi_1(\bar{c}) = \tfrac{2}{3}\,\bar{c}^3 + \tfrac{5}{3}\,\bar{c} + (\tfrac{2}{3}\,\bar{c}^4 + \tfrac{4}{3}\bar{c}^2 + \tfrac{11}{6})\; Z(\bar{c}) \;,$$

$$\Phi_2(\bar{c}) = -\tfrac{4}{3}\,\bar{c}^2 - 2 - \tfrac{4}{3}\bar{c}^3 Z(\bar{c}) + (\tfrac{4}{3}\bar{c}^2 + \tfrac{2}{3})\, Z^2(\bar{c}) \;,$$

$$\Phi_3(\bar{c}) = \tfrac{2}{3}\,\bar{c} + (\tfrac{2}{3}\bar{c}^2 - \tfrac{5}{3})\, Z(\bar{c}) - \tfrac{4}{3}\bar{c} Z^2(\bar{c}) \;. \tag{5.53}$$

Just as in section 2, we can apply the winding theorem to find the number of zeros of F in the upper and lower half $\bar{c}$-plane. We find that for no positive real value of k (i.e. negative imaginary ϵ) is there a zero of F in the lower half plane, while for sufficiently small k there are three zeros in the upper half plane, one pure imaginary, corresponding to the heat conduction mode, and two with equal imaginary parts and opposite real parts, corresponding to the sound modes. As k increases, these zeros move rapidly toward the imaginary axis, the two sound modes arriving first, when $k_c = 1.853\, n\sigma\gamma$ followed soon after by the heat mode, which arrives when $k_c = 1.918\, n\sigma\gamma$. For still larger values of k there are no zeros of F.

To see more in detail the paths of the zeros in the upper half plane we consider first the case of small ϵ, i.e., long wavelengths. Using the asymptotic expansion (5.19) we can, after considerable effort, obtain the following expansions of the three zeros in powers of ϵ:

$$\bar{c} = \frac{\gamma}{\epsilon} + \frac{\epsilon}{2\gamma} + \frac{9\epsilon^3}{20\gamma^3} + \ldots \;, \tag{5.54a}$$

$$\bar{c} = \frac{\gamma}{\epsilon} + \frac{\epsilon}{2\gamma} + \frac{31\epsilon^3}{90\,\gamma^3} + \ldots$$

$$\pm\, (\tfrac{5}{6})^{\frac{1}{2}} \left[1 - \frac{7\epsilon^2}{60\gamma^2} - \frac{281\,\epsilon^4}{800\,\gamma^4} + \ldots \right] \;. \tag{5.54b}$$

For still larger values of k the zeros of (5.52) must be determined numerically. We have done so and the path of the zeros is shown in fig.17.

The pure imaginary root corresponds to the heat conduction mode. As we have done in earlier discussions, we introduce the dimensionless frequency and wave number according to (4.48)

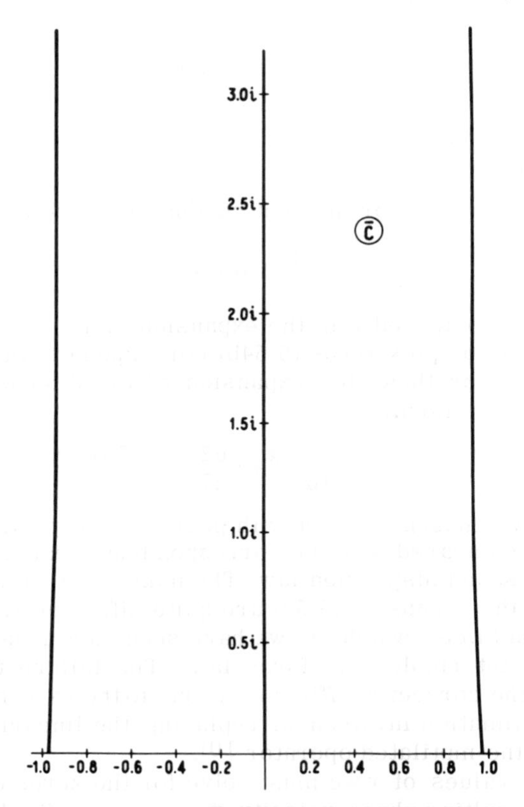

Fig. 17. Path of the zeros in the $\bar{c}$-plane for the three-dimensional model.

with [9])

$$\eta = \frac{m}{\sigma\gamma} \left(\frac{kT}{2m}\right)^{\frac{1}{2}} . \tag{5.55}$$

Hence, we put

$$\xi = \frac{3}{5n\sigma\gamma} \left(\frac{m}{2kT}\right)^{\frac{1}{2}} \omega ,$$

$$x = \left(\tfrac{3}{10}\right)^{\frac{1}{2}} \frac{k}{n\sigma\gamma} , \tag{5.56}$$

or, referring to (5.44) and (5.45),

$$E = -i\,\frac{5\gamma}{3}\,\xi\,, \qquad \epsilon = -i\,(\tfrac{10}{3})^{\frac{1}{2}}\,\gamma x\,,$$

$$\bar{c} = (\tfrac{5}{6})^{\frac{1}{2}}\,\frac{\xi + i\frac{3}{5}}{x}\,. \tag{5.57}$$

The expansion (5.54a) for the heat conduction mode becomes

$$\xi = -\,i\,(x^2 - 3x^4 + \dots)\,, \tag{5.58}$$

which is to be compared with the expansion (2.49).

The pair of complex zeros (5.54b) correspond to the longitudinal sound modes. For these the expansion of the dimensionless wave number takes the form:

$$\xi = x - i\,x^2 + \frac{7}{18}\,x^3 + i\,\frac{62}{27}\,x^4 - \frac{281}{72}\,x^5 + \dots\,, \tag{5.59}$$

where we have chosen the root with positive real part. This expansion is to be compared with the corresponding expansions (2.50) and (4.83) of the sound dispersion law. The numerical values of the coefficients in the expansion (5.59) are quite different from those in the expansion (4.83), which, as we have seen, are relatively insensitive to the intermolecular force law. The failure to get somewhere near the correct coefficients is due to the extreme character of the approximation involved in replacing the linearized collision operator by the mutilated operator [10].

For large values of x we must solve for the zeros of F numerically. The resulting phase velocity is shown in fig. 18 where the quantity:

$$\frac{v_{\mathrm{ph}}}{v_{\mathrm{o}}} = \mathrm{Re}\left\{\frac{\xi}{x}\right\}\,, \tag{5.60}$$

is plotted as a function of x. Here $v_{\mathrm{o}} = (5kT/3m)^{\frac{1}{2}}$ is the Laplace sound speed. Similarly in fig. 19 we plot the quantity:

$$1/kv_{\mathrm{o}}\tau = -\,\mathrm{Im}\left\{\frac{\xi}{x}\right\}\,, \tag{5.61}$$

where τ is the relaxation time, as defined in (1.5). These curves as well as the coefficients in the expansions (5.58) and (5.59) have, of course, only a qualitative significance. The mutilated operator is not the correct linearized Boltzmann collision operator!

We close this section with some remarks about the applications

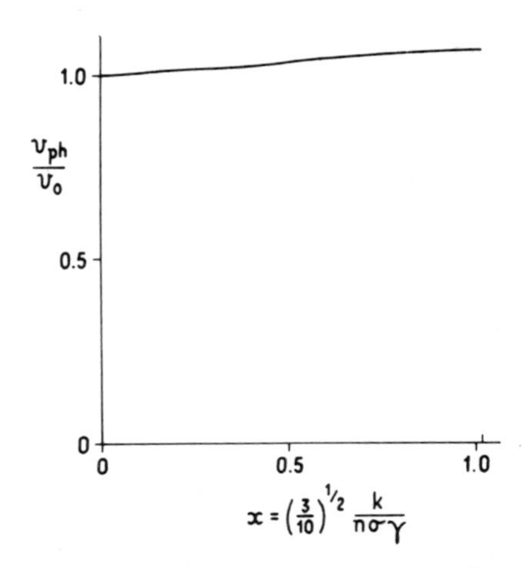

Fig. 18. The phase velocity as a function of dimensionless wave number for the three-dimensional model.

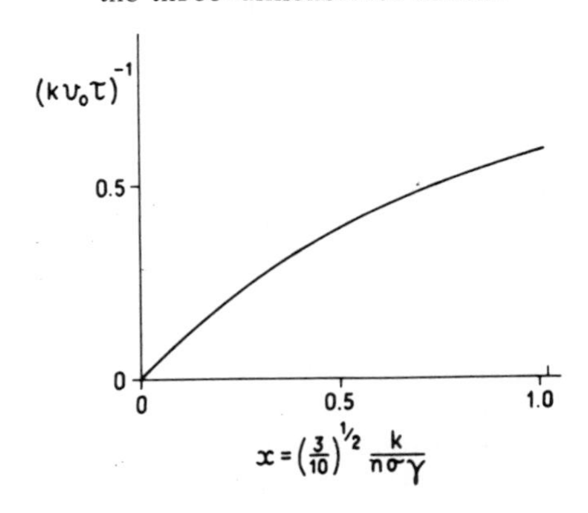

Fig. 19. The relaxation time as a function of dimensionless wave number for the three-dimensionless model.

of more general model equations which have been made by a number of authors. The general idea is as indicated in section 1, the mutilation of the collision operator is generalized to include additional non-zero discrete eigenvalues. Perhaps the most extensive calculation to date is that of Sirovich and Thurber [11], who chose a mutilated collision operator of the form:

$$\boldsymbol{J}h = -\gamma h + \gamma \sum_{(r,l)} (\psi_{rl}, h)\, \psi_{rl} + \sum_{(r,r',l)} J^{l}_{rr'} (\psi_{rl}, h) \psi_{r'l}\,, \tag{5.62}$$

where the ψ_{rl} are normalized Burnett functions and $J^{l}_{rr'}$ is the matrix element of the exact collision operator introduced in chapter III. (For simplicity the discussion is restricted to axially symmetric solutions). The sums in (5.62) are finite sums over a selected set of the labels (r,l). If the set is chosen to be (0,0), (0,1), (2,0), corresponding to the three axially symmetric eigenfunctions of $\boldsymbol{J}$ with eigenvalue zero, we get exactly the mutilated collision operator (5.1), which Sirovich and Thurber call the 3-moment model. Adding (0,2) and (1,1) gives a 5-moment model; adding to these (0,3), (2,0) and (1,2) gives an 8-moment model, adding further (0,4), (2,1), (1,3) gives an 11-moment model, and so on. Note that these successive models correspond exactly to the successive truncations of the Chang determinant, discussed in chapter IV. The discussion of these higher model equations goes exactly as in our discussion of the 3-moment model; it is only more complicated. Sirovich and Thurber have discussed the 3-, 8- and 11-moment models numerically for the case of Maxwell molecules and the case of hard spheres. They find a rather good agreement with experiment.

There is, however, an objection to this method of model equations which in our opinion is rather serious. This is that as one adds more and more moments, the mutilated operator (5.62) does *not* converge to the linearized collision operator, whose matrix elements are $J^{l}_{rr'}$. This is due to the fact that the linearized collision operator is an unbounded operator, so the matrix elements increase without limit. The model equations undoubtedly give a correct qualitative picture of the dispersion law, but if the aim is to test the Boltzmann equation by comparison with experimental measurements of sound dispersion and absorption, one must have reliable *quantitative* solutions of this equation. At our present state of understanding of the mathematical problems associated with the solution of the linearized Boltzmann equation, the method of model equations remains an uncontrolled approximation method.

NOTES

1) E. P. Gross and E. A. Jackson, Phys. Fluids 2 (1959) 432. The mutilated collision operator (5.1) is sometimes called the "Krook operator" since its use was first suggested in a paper by P. F. Bhatnager, E. P. Gross and M. Krook, Phys. Rev. 94 (1954) 511.

The successive approximation scheme as proposed by M. Kac

was based on the assumption that the spectrum of the collision operator consisted only of discrete eigenvalues of finite multiplicity which tend to a lower limit, $-\gamma$. If this were the case, one could write

$$\boldsymbol{J} h = -\gamma h + \boldsymbol{K} h \ ,$$

and the operator $\boldsymbol{K}$ would be a bounded operator with a spectrum of discrete eigenvalues of finite multiplicity which tend to zero as a limit point. In other words, $\boldsymbol{K}$ would be a completely continuous operator. For such operators there is a well-known theorem which states that they may be successively approximated by a sequence of operators of finite rank. (See, e.g., F. Riesz and B. Sz.-Nagy, *Functional Analysis* (Frederick Ungar Publishing Co., New York, 1955) Section 85. This is exactly the Gross-Jackson Kac Scheme. However, we now know that the collision operator is unbounded and, with the exception of the case of Maxwell molecules, perhaps has a spectrum which is in part continuous. For such operators the Gross-Jackson-Kac scheme does not converge.

2) This model equation is inspired by a one-dimensional Boltzmann-like equation proposed by M. Kac, *Proceedings of the Third Berkeley Symposium on Mathematics, Statistics and Probability* (University of California Press, Berkeley, 1956) Vol. 3, pp. 171-197. Kac's equation is

$$\frac{\partial f}{\partial t} + c\frac{\partial f}{\partial x} = \int_{-\infty}^{\infty} \mathrm{d}c_1 \int_{-\pi}^{\pi} \mathrm{d}\theta\, F(\theta)\,[f'f_1' - ff_1] \ ,$$

where $c' = c\cos\theta + c_1 \sin\theta$ and $c_1' = -c\sin\theta + c_1\cos\theta$. Here $F(\theta)$ is the collision probability; the Kac equation is related to the case of Maxwell molecules for which the collision probability is independent of relative velocity. Since $c'^2 + c_1'^2 = c^2 + c_1^2$, the additive constants of the motion of a binary collision are 1 and c^2. If we linearize by writing:

$$f = \pi^{-\frac{1}{2}}\, \mathrm{e}^{-c^2}\,[1 + h(x,c,t)] \ ,$$

and neglecting second order quantities in h, we get:

$$\frac{\partial h}{\partial t} + c\frac{\partial h}{\partial x} = \boldsymbol{L} h \ ,$$

where

$$\boldsymbol{L} h = \pi^{-\frac{1}{2}} \int_{-\infty}^{\infty} \mathrm{d}c_1\, \mathrm{e}^{-c_1^2} \int_{-\pi}^{\pi} \mathrm{d}\theta\, F(\theta)\,[h' + h_1' - h - h_1] \ .$$

The eigenfunctions of this operator are the Hermite polynomials, $H_n(c)$, with eigenvalue

$$\lambda_n = \int_{-\pi}^{\pi} d\theta\, F(\theta)\,[\cos^n\theta + (-)^n \sin^n\theta - 1 - \delta_{n,0}] \;.$$

The eigenvalues λ_0 and λ_2 are zero, corresponding to the conserved quantities 1 and c^2. All other eigenvalues are negative. When we mutilate by choosing $\lambda_n = -\gamma$, $n \geqslant 2$ we get, using the completeness of the eigenfunctions,

$$\boldsymbol{L}h = -\gamma h + \frac{\gamma}{\pi^{\frac{1}{2}}} \int_{-\infty}^{\infty} dc_1\, e^{-c_1^2}[1 + 2(c^2 - \tfrac{1}{2})(c_1^2 - \tfrac{1}{2})]h(c_1) \;.$$

However, this operator does not lead to sound-like propagating modes as does the operator (5.3).

3) The following discussion of the solution of equation (5.6) is inspired by similar discussions for equations occurring in the theories of neutron transport and plasma kinetic equations. See, e.g., K. M. Case and P. F. Zweifel, *Linear Transport Theory* (Addison-Wesley Publishing Co., Reading, Mass., 1967) esp. chapter 4 and chapter 10. See also N. G. van Kampen, *Physica* 21 (1955) 949.

4) See, e.g., E. T. Whittaker, *Modern Analysis*, 4th ed. (Cambridge University Press, 1952) p. 119.

5) If we differentiate (5.18) with respect to k we find

$$k\,\frac{d\bar{c}}{dk} = -\frac{\frac{k}{\gamma}[2\bar{c} + (2\bar{c}^2 + 1)\,Z(\bar{c})] + 4[1 + \bar{c}Z(\bar{c})]}{2\left(1 + i\frac{k\bar{c}}{\gamma}\right)[2\bar{c} + (2\bar{c}^2 - 1)\,Z(\bar{c})]} \;, \qquad \text{(a)}$$

where we have used (C.3). If now we solve (5.18) for $Z(\bar{c})$ in terms of $\bar{c}$ and k, and then substitute the resulting expression in (a), we get the following differential equation for the path of the zeros:

$$\frac{d\bar{c}}{dk} = -i\,\frac{2\gamma^2 + 4i\gamma\, k\bar{c} - k^2(2\bar{c}^2 + 1)}{2(\gamma + k\bar{c})[2\gamma^2 + 4i\gamma\, k\bar{c} - k^2(2\bar{c}^2 - 1)} \;. \qquad \text{(b)}$$

In principle we can use this result to find the path of the zeros without recourse to tables of Z.

6) If we use (5.8) and put $\omega = i\gamma E$, $k = i\gamma\epsilon$, then upon multiplying each row of the determinant by ϵ we get a determinant very similar in form to the Chang determinant given on page 158.

7) See e.g. H. Jeffreys, *Asymptotic Expansions* (Clarendon Press, Oxford, 1962) pp. 99-103.

8) For example, inserting (5.48) in (5.43a) and using the integral identities:

$$\pi^{-\frac{3}{2}} \int \mathrm{d}\boldsymbol{c}\, \mathrm{e}^{-c^2} \frac{\boldsymbol{c}\cdot\boldsymbol{\varphi}}{\hat{\boldsymbol{k}}\cdot\boldsymbol{c} - \bar{c}}$$

$$= \pi^{-\frac{3}{2}} \int_{-\infty}^{\infty} \mathrm{d}c_x \int_{-\infty}^{\infty} \mathrm{d}c_y \int_{-\infty}^{\infty} \mathrm{d}c_z \; \mathrm{e}^{-c_x^2 - c_y^2 - c_z^2} \frac{c_x\varphi_x + c_y\varphi_y + c_z\varphi_z}{c_z - \bar{c}}$$

$$= \pi^{-\frac{1}{2}} \int_{-\infty}^{\infty} \mathrm{d}c_z \; \mathrm{e}^{-c_z^2} \frac{c_z\varphi_z}{c_z - \bar{c}}$$

$$= [1 + \bar{c}Z(\bar{c})]\hat{\boldsymbol{k}}\cdot\boldsymbol{\varphi} \;,$$

and

$$\pi^{-\frac{3}{2}} \int \mathrm{d}\boldsymbol{c}\; \mathrm{e}^{-c^2} \frac{c^2 - \frac{3}{2}}{\hat{\boldsymbol{k}}\cdot\boldsymbol{c} - \bar{c}} = \pi^{-\frac{1}{2}} \int_{-\infty}^{\infty} \mathrm{d}c_z \; \mathrm{e}^{-c_z^2} \frac{c_z^2 - \frac{1}{2}}{c_z - \bar{c}} = \bar{c} + (\bar{c}^2 - \tfrac{1}{2})\, Z(\bar{c}) \;,$$

we get (5.49a).

9) We get this result from (4.70), with

$$a_{02}^{(1)} = 1/3^{\frac{1}{2}}\gamma \;,$$

which in turn we get by using the mutilated collision operator (5.1) in (4.28b).

10) This is exemplified by the Eucken ratio

$$f = \kappa / c_v \eta \;,$$

which, as we have remarked, both experimentally and theoretically is found to have a value very close to $\frac{5}{2}$ for all monoatomic gases. If we use the mutilated operator (5.1) to determine κ and η from (4.70) we find $f = \frac{5}{3}$, which is far from the correct value. This is the reason for the discrepancy in the coefficient of the x^2 term, the "Kirchhoff" absorption, in the expansion (5.59).

11) See L. Sirovich and J. Thurber, J. Acoust. Soc. Amer. 37 (1965) 329.

Chapter VI

OUTLOOK

1. EXPERIMENT

In our opinion the principal reason for experimental and theoretical studies of sound dispersion in monoatomic gases is to improve our knowledge and understanding of kinetic theory, in particular to test experimentally the Boltzmann equation. At low frequencies, i.e., when the dimensionless frequency:

$$\xi = \omega\eta/\rho v_0^2 \,, \tag{6.1}$$

is small, the agreement between theory and experiment is satisfactory, as we showed in chapter IV. There it is apparent that the theory is in advance of experiment. That is, by the perturbation method we can compute the first few coefficients in the expansion of the phase velocity and absorption coefficient in powers of ξ to any desired accuracy. We might just remark here that all the numerical results in chapter IV were obtained by hand computation; even modest use of an electronic computer will allow us to do much more. However, it would be desirable to have experiments sufficiently accurate to distinguish between different molecular models, which the present experiments cannot do. In addition it would be of interest to measure the coefficients as functions of temperature, e.g., to attempt experimentally to measure the correction factor Δ_1 to the dispersion, as given in fig. 5. Needless to say, this is quite an order! [1].

At higher frequencies, when ξ is not small, the situation is reversed. In fig. 20 is shown the observed phase velocity and absorption coefficient for ξ up to about unity [2]. Not shown are a few additional experimental points between $\xi = 1$ and $\xi = 2$, which indicate $\alpha v_0/\omega$ remains practically constant, while v_0/v_{ph} continues to decrease gradually to about 0.53 at $\xi = 2$. The curves labelled Navier-Stokes and 3-moment model are, respectively, obtained from the Navier-Stokes dispersion law (2.35) and the dispersion law (5.52) for the model equation discussed in chapter V. Not shown are

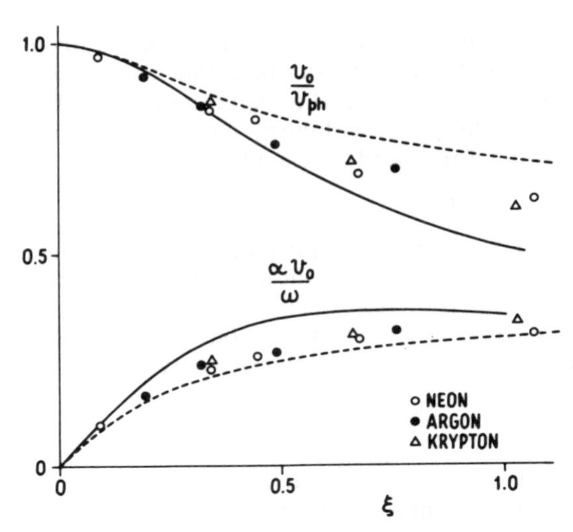

Fig. 20. The dispersion and absorption of sound as functions of the dimensionless frequency. The experimental points are those of Greenspan. The solid curves are the predicted results from the Navier-Stokes equation. The broken curves are the predicted results from the three-dimensional model equation of chapter V.

theoretical curves based on higher moment equations, which are in somewhat better agreement with experiment [3].

At still higher frequencies, when ξ is large, one is in the Knudsen regime where the wave length is short compared with the mean free path. As indicated in our discussion of the model equations, the sound mode with its associated dispersion law no longer exists in this regime. Instead, the disturbance is a superposition of a continuum of Knudsen modes with no dispersion law relating frequency and wave length. The particular combination of these modes which occurs and, hence, the *apparent* dispersion law is determined by the boundary conditions at the source. This has been shown explicitly by Kahn and Mintzer who consider sound propagation in the Knudsen regime [4]. They obtain rather good agreement with experiment using a simple boundary condition (so called perfect accomodation) and it is clear from their discussion that the agreement can be improved by modifying the boundary condition.

Before we leave this discussion of the experiments, we wish to stress the important constraint that L, the distance from the source to the detector, be much greater than the mean free path Λ. This is an obviously necessary constraint if the sound propagation is to be governed solely by collisions *within* the gas, i.e., by the Boltzmann equation. If one relaxes this condition, then one speaks of free molecular propagation rather than sound propagation, and it is found experimentally, for example, that the measured phase velocity depends upon L. From the experimentally observed onset of L-dependence of the phase velocity, one can conclude that a necessary constraint for sound propagation is [5]

$$L \geqslant 15\,\Lambda \ . \tag{6.2}$$

This constraint makes sound propagation experiments more and more difficult for higher frequencies, and experimental uncertainties correspondingly greater. The amplitude of the sound wave at the detector is less than that at the source by a factor $e^{-\alpha L}$, where α is the absorption coefficient. But αL may be written

$$\alpha L = \frac{\alpha v_0}{\omega}\,\xi\,\frac{L}{\Lambda} \ . \tag{6.3}$$

Thus, for $\xi = 1$, from fig. 20, we see $\alpha v_0/\omega \approx 0.3$ and, using (6.2) we have $\alpha L \approx 4.5$. Hence the amplitude is diminished by $e^{-4.5} \approx \frac{1}{90}$. Since $\alpha v_0/\omega$ is practically constant between $\xi = 1$ and $\xi = 2$, an increase of ξ from 1 to 2 is accompanied by a further decrease of two orders of magnitude in amplitude. This illustration shows how much more difficult the sound propagation experiments become at higher frequencies.

2. THEORY

The outstanding problem relating to sound dispersion at intermediate frequencies is that of finding a satisfactory method of solution of the linearized Boltzmann equation in this regime. We wish to outline in this section a proposed method which can lead to such a solution, but, as we shall see, with considerable difficulty [6].

We consider the case of intermolecular potentials for which the separation (3.30) of the linearized collision operator can be made. This includes the case of elastic spheres and also a more realistic potential, such as the Lennard-Jones potential, provided the correct quantum mechanical cross section is used. In this case we can write the linearized Boltzmann equation (4.4) in the form:

$$[E - \epsilon c_z + m(c)]h_{\omega,k} = \boldsymbol{K} h_{\omega,k} \ . \tag{6.4}$$

As in our discussion of the model equations, we distinguish two cases, depending upon whether the quantity:

$$E - \epsilon\, c_z + m(c) \ , \tag{6.5}$$

vanishes for some value of c. For simplicity in the discussion we consider the case of given wave vector k, i.e., ϵ, given by (4.5), is a purely imaginary constant. Then using the example of hard spheres, where $m(c)$ is given by (3.53), the quantity (6.5) vanishes for E in the shaded region of the E-plane shown in fig. 21, bounded by the curve:

$$E = m(c) \pm \epsilon c \ , \qquad 0 < c < \infty \ . \tag{6.6}$$

For E in this region, we can "solve" (6.4) for $h_{\omega,k}$ in the form:

$$h_{\omega,k} = A_{\omega,k}\, \delta[\,|E - \epsilon c_z + m(c)|\,] + P\, \frac{1}{E - \epsilon c_z + m(c)}\, \boldsymbol{K} h_{\omega,k} \ . \tag{6.7}$$

This is an inhomogeneous singular integral equation for $h_{\omega,k}$ and we can expect that in general solutions will exist. These solutions will be distributions, not functions, as we see if we note that for $h_{\omega,k}$ a linear combination of a δ-function and a principal value function, $\boldsymbol{K}h_{\omega,k}$ is a function. This follows from the fact that the integral operator $\boldsymbol{K}$ is completely continuous. Clearly these solutions, are analogous to the Knudsen mode solutions of the model equations discussed in chapter V, the difference being that since $m(c)$ is velocity dependent, there is a region instead of a line in the E-plane for which these solutions exist. Here, as in our discussion of the model equation, we interpret these modes, for which there is no dispersion law, as single particle or Knudsen modes.

The remaining case, where E is outside the shaded region and

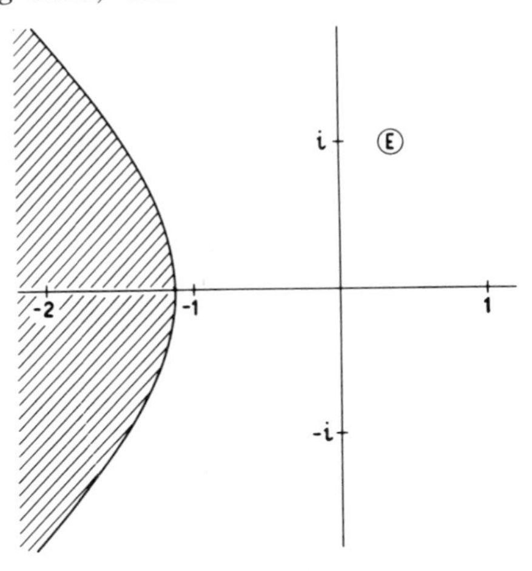

Fig. 21.

the quantity (6.5) does not vanish, is that of greatest interest for this includes the sound modes. In this case, we can rewrite (6.4) in the form:

$$\left[1 - \frac{1}{E - \epsilon c_z + m(c)} \boldsymbol{K}\right] h_{\omega,k} = 0 \ . \tag{6.8}$$

This is a homogeneous integral equation for the determination of $h_{\omega,k}$, and in general solutions will exist only for discrete values of E; those discrete values growing out of $E = 0$, $\epsilon = 0$, will be the dispersion laws for the sound and heat conduction modes [7].

The solutions of (6.8) will be square integrable, so it makes sense to seek a solution expanded in terms of the Burnett functions:

$$h_{\omega,k} = \sum_{r=0}^{\infty} \sum_{l=0}^{\infty} a_{rl} \psi_{rl} \ . \tag{6.9}$$

Putting this into (6.8) and forming the scalar product with ψ_{rl}, we get an infinite set of linear homogeneous equations for the determination of the coefficients a_{rl}, the general equation being

$$a_{rl} - \sum_{r'=0}^{\infty} \sum_{l'=0}^{\infty} \left(\psi_{rl} \, , \frac{1}{E - \epsilon c_z + m(c)} \boldsymbol{K} \psi_{r'l'}\right) a_{r'l'} = 0 \ . \tag{6.10}$$

In order that there be a non-trivial solution of these equations, it is necessary that the associated infinite determinant must vanish, and setting this determinant equal to zero gives us the desired dispersions law relating E and ϵ (i.e., relating ω and k).

The analogy of this method with the method of Mott-Smith, Chang and Uhlenbeck discussed in chapter IV is obvious. The only difference is that we divide by the factor (6.5) before forming the determinant. But this is an essential difference, since by doing so we have excluded the Knudsen modes, which cannot be expanded in the form (6.9). Continuing, we would propose to approximate this infinite determinant by successive truncations, exactly as with the Chang determinant. The first truncation would be the 3×3 determinant

$$\begin{vmatrix} 1 - \left(\psi_{00}, \frac{m}{E - \epsilon c_z + m} \psi_{00}\right) & -\left(\psi_{01}, \frac{m}{E - \epsilon c_z + m} \psi_{00}\right) & -\left(\psi_{10}, \frac{m}{E - \epsilon c_z + m} \psi_{00}\right) \\ -\left(\psi_{00}, \frac{m}{E - \epsilon c_z + m} \psi_{01}\right) & 1 - \left(\psi_{01}, \frac{m}{E - \epsilon c_z + m} \psi_{01}\right) & -\left(\psi_{10}, \frac{m}{E - \epsilon c_z + m} \psi_{01}\right) \\ -\left(\psi_{00}, \frac{m}{E - \epsilon c_z + m} \psi_{10}\right) & -\left(\psi_{01}, \frac{m}{E - \epsilon c_z + m} \psi_{10}\right) & 1 - \left(\psi_{10}, \frac{m}{E - \epsilon c_z + m} \psi_{10}\right) \end{vmatrix} \ . \tag{6.11}$$

Here we have used the fact that:

$$K\psi_{rl} = m(c)\,\psi_{rl} \;, \tag{6.12}$$

for ψ_{00}, ψ_{01} and ψ_{10}, i.e., for the functions corresponding to the zero eigenvalue of $\boldsymbol{J}$. In the next approximation one would add $\psi_{0,2}$ and $\psi_{1,1}$ to obtain a five by five determinant, and so on.

We should emphasize that what we have outlined here is a *program* not a solution. The essential difficulty is that even the simplest matrix elements which appear, for example:

$$\left(\psi_{00}, \frac{m}{E - \epsilon c_z + m}\,\psi_{00}\right) = \pi^{-\frac{3}{2}} \int \mathrm{d}\boldsymbol{c}\, \mathrm{e}^{-c^2} \frac{m(c)}{E - \epsilon c_z + m(c)} \;, \tag{6.13}$$

are integrals which must be evaluated numerically. Obviously a great deal of work with a high-speed computer will be involved.

If $m(c)$ is replaced by a constant, say $m(0)$, then the integrals which appear in the 3×3 determinant (6.11) can all be expressed in terms of the function $Z(z)$, discussed in appendix C. In fact, with this replacement, the determinant (6.11) becomes exactly the dispersion law (5.52) for the 3-moment model equation, with $\gamma = m(0)$. This relation with the model equations does not hold in higher approximation.

We will say no more about this method for solution of the linearized Boltzmann equation in the intermediate range of frequencies, since it is only a program. Our main point is that the next step in the theoretical development is that this or some other method of solution should be carried to the point of producing a satisfactory *quantitative* theoretical dispersion law for comparison with the experimental data shown in fig. 20.

NOTES

1) We should remark here that experiments measuring the Brillouin scattering from gases, which in principle test the coefficients in the inverse expansion of ξ in powers of $x = k\eta/\rho v_0$, the dimensionless wave number, need be only about $\frac{1}{10}$ as precise as the sound dispersion measurements to determine Δ_1 with the same accuracy. This is because of the small numerical factor multiplying Δ_1 in the expansion (4.84) as compared with the expansion (4.83). For an account of the Brillouin scattering from a hydrodynamical point of view, see, e.g., R. D. Mountain, Rev. Mod. Phys. 38 (1966) 205 and also J. Foch, Phys. Fluids 11 (1968) 2336.

2) Greenspan, *op. cit.*, note 2, chapter I. Meyer and G. Sessler, Z. Physik 149 (1957) 15, investigated sound propagation in Argon over essentially the same range of ξ values as Greenspan and found good agreement with his results.

3) See Sirovich and Thurber, *op. cit.*, note 11, chapter V.

4) D. Kahn and D. Mintzer, Phys. Fluids 8 (1965) 1090. The role of the boundary conditions at the source in the Knudsen regime suggests an interpretation of the critical wave number k_c introduced in chapter V. We would say that k_c separates the long wavelength regime, where the sound dispersion law is determined by the properties of the medium, from the Knudsen regime, where the apparent dispersion relation is determined by the boundary conditions at the source.

5) See G. Sessler, J. Acoust. Soc. Amer. 38 (1965) 974, where however a slightly different definition of the mean free path is used. A comparable constraint was imposed by Greenspan. See M. Greenspan and M. C. Thompson Jr., J. Acoust. Soc. Amer. 25 (1953) 92, esp. section 4 and (for Greenspan's definition of the mean free path) M. Greenspan, J. Acoust. Soc. Amer. 22 (1950) 568.

6) A similar program has been outlined by G E. Skvortsov, Soviet Phys. JETP 25 (1967) 853.

7) The point here is that $\boldsymbol{K}$ is completely continuous (we consider the case of hard spheres) and $[E - \epsilon c_z + m(c)]^{-1}$, considered as an operator, is bounded for E outside the shaded region in fig. 21. But the product of a bounded operator and a completely continuous operator is completely continuous. (See F. Riesz and B. Sz.-Nagy, *Functional Analysis* (Frederick Ungar Publishing Co., New York, 1955, section 76). Hence, a solution of (6.8) is an eigenfunction of a completely continuous operator with eigenvalue unity. But a completely continuous operator has only discrete non-zero eigenvalues of finite multiplicity with square integrable eigenfunctions.

APPENDIX A

The matrix elements of $\boldsymbol{J}$ *in terms of the* Ω*-integrals.*

In this appendix we give expressions for the matrix elements of $\boldsymbol{J}$ in terms of the Ω-integrals of Chapman and Cowling:

$$\Omega^{(j)}(l) = 2^{-l-\frac{3}{2}}\, \pi^{\frac{1}{2}} \int_0^{\infty} \mathrm{d}g\, \mathrm{e}^{-\frac{1}{2}g^2}\, g^{2l+2}\, \phi^{(j)}(g) \ ,$$

where

$$\phi^{(j)}(g) = \frac{\sigma}{2\pi}\left(\frac{2\,kT}{m}\right)^{\frac{1}{2}} \int \mathrm{d}\Omega\, F(g,\theta)(1-\cos^{j}\theta) \ .$$

Here $F(g,\theta)$ is given by (3.18).

For elastic spheres $F(g,\theta) = ga^2/4\sigma$, and

$$\Omega^{(2j)}(l) = a^2\left(\frac{\pi\, kT}{m}\right)^{\frac{1}{2}} \frac{j}{2j+1}\,(l+1)! \ ,$$

while for power law forces:

$$F(g,\theta) = \frac{a^2}{\sigma}\left(\frac{2\,\varphi_s}{kT}\right)^{\frac{2}{s}} g^{1-\frac{4}{s}}\, f_s(\theta) \ ,$$

and

$$\Omega^{(2j)}(l) = a^2\left(\frac{kT}{m}\right)^{\frac{1}{2}-\frac{2}{s}} \left(\frac{\varphi_s}{m}\right)^{\frac{2}{s}} \frac{\Gamma(l+2-\frac{2}{s})}{2\,\Gamma(\frac{1}{2})} \int \mathrm{d}\Omega\, f_s(\theta)(1-\cos^{2j}\theta) \ .$$

The expressions for the first few matrix elements $J^{l}_{\gamma\gamma'}$ of the linearized collision operator in terms of these integrals follow.

$$\sigma\left(\frac{2\,kT}{m}\right)^{\frac{1}{2}} J^{0}_{22} = -\frac{16}{15}\,\Omega^{(2)}(2) \ ,$$

$$\sigma\left(\frac{2\,kT}{m}\right)^{\frac{1}{2}} J^{0}_{23} = -\left(\frac{2}{21}\right)^{\frac{1}{2}} \left[\frac{28}{5}\,\Omega^{(2)}(2) - \frac{8}{5}\,\Omega^{(2)}(3)\right] ,$$

$$\sigma\left(\frac{2\,kT}{m}\right)^{\frac{1}{2}} J^{0}_{33} = -\left[\frac{18}{5}\,\Omega^{(2)}(2) - \frac{8}{5}\,\Omega^{(2)}(3) + \frac{8}{35}\,\Omega^{(2)}(4)\right] ,$$

$$\sigma\left(\frac{2\,kT}{m}\right)^{\frac{1}{2}} J^{0}_{24} = -\left(\frac{3}{7}\right)^{\frac{1}{2}} \left[\frac{14}{5}\,\Omega^{(2)}(2) - \frac{8}{5}\,\Omega^{(2)}(3) + \frac{8}{45}\,\Omega^{(2)}(4)\right] ,$$

$$\sigma\left(\frac{2kT}{m}\right)^{\frac{1}{2}} J^{0}_{34} = -2^{\frac{1}{2}}\left[\frac{33}{10}\,\Omega^{(2)}(2) - \frac{15}{7}\,\Omega^{(2)}(3) + \frac{10}{21}\,\Omega^{(2)}(4) - \frac{4}{105}\,\Omega^{(2)}(5)\right.$$

$$\sigma\left(\frac{2kT}{m}\right)^{\frac{1}{2}} J^{0}_{44} = -\left[\frac{143}{20}\,\Omega^{(2)}(2) - \frac{594}{35}\,\Omega^{(2)}(3) + \frac{538}{315}\,\Omega^{(2)}(4) - \frac{8}{45}\,\Omega^{(2)}(5)\right.$$

$$\left. + \frac{4}{315}\,\Omega^{(2)}(6) + \frac{16}{945}\,\Omega^{(4)}(4)\right] ,$$

$$\sigma\left(\frac{2kT}{m}\right)^{\frac{1}{2}} J^{1}_{11} = -\frac{16}{15}\,\Omega^{(2)}(2) ,$$

$$\sigma\left(\frac{2kT}{m}\right)^{\frac{1}{2}} J^{1}_{12} = -7^{\frac{1}{2}}\left[\frac{8}{15}\,\Omega^{(2)}(2) - \frac{16}{105}\,\Omega^{(2)}(3)\right] ,$$

$$\sigma\left(\frac{2kT}{m}\right)^{\frac{1}{2}} J^{1}_{22} = -\left[\frac{44}{15}\,\Omega^{(2)}(2) - \frac{16}{15}\,\Omega^{(2)}(3) + \frac{16}{105}\,\Omega^{(2)}(4)\right] ,$$

$$\sigma\left(\frac{2kT}{m}\right)^{\frac{1}{2}} J^{1}_{13} = -\left(\frac{2}{21}\right)^{\frac{1}{2}}\left[\frac{21}{5}\,\Omega^{(2)}(2) - \frac{12}{5}\,\Omega^{(2)}(3) + \frac{4}{15}\,\Omega^{(2)}(4)\right] ,$$

$$\sigma\left(\frac{2kT}{m}\right)^{\frac{1}{2}} J^{1}_{23} = -\left(\frac{2}{3}\right)^{\frac{1}{2}}\left[\frac{9}{2}\,\Omega^{(2)}(2) - \frac{87}{35}\,\Omega^{(2)}(3) + \frac{10}{21}\,\Omega^{(2)}(4) - \frac{4}{105}\,\Omega^{(2)}(5)\right.$$

$$\sigma\left(\frac{2kT}{m}\right)^{\frac{1}{2}} J^{1}_{33} = -\left[\frac{231}{5}\,\Omega^{(2)}(2) - \frac{27}{7}\,\Omega^{(2)}(3) + \frac{313}{315}\,\Omega^{(2)}(4) - \frac{4}{35}\,\Omega^{(2)}(5)\right.$$

$$\left. + \frac{2}{315}\,\Omega^{(2)}(6) + \frac{16}{945}\,\Omega^{(4)}(4)\right] ,$$

$$\sigma\left(\frac{2kT}{m}\right)^{\frac{1}{2}} J^{2}_{00} = -\frac{8}{5}\,\Omega^{(2)}(2) ,$$

$$\sigma\left(\frac{2kT}{m}\right)^{\frac{1}{2}} J^{2}_{01} = -\left(\frac{2}{7}\right)^{\frac{1}{2}}\left[\frac{14}{5}\,\Omega^{(2)}(2) - \frac{4}{5}\,\Omega^{(2)}(3)\right] ,$$

$$\sigma\left(\frac{2kT}{m}\right)^{\frac{1}{2}} J^{2}_{11} = -\left[\frac{43}{15}\,\Omega^{(2)}(2) - \frac{4}{5}\,\Omega^{(2)}(3) + \frac{4}{35}\,\Omega^{(2)}(4)\right] ,$$

$$\sigma\left(\frac{2kT}{m}\right)^{\frac{1}{2}} J^{2}_{02} = -\left(\frac{2}{7}\right)^{\frac{1}{2}}\left[\frac{21}{10}\Omega^{(2)}(2) - \frac{6}{5}\Omega^{(2)}(3) + \frac{2}{15}\Omega^{(2)}(4)\right] ,$$

$$\sigma\left(\frac{2kT}{m}\right)^{\frac{1}{2}} J^{2}_{12} = -\left[\frac{13}{4}\Omega^{(2)}(2) - \frac{107}{70}\Omega^{(2)}(3) + \frac{5}{21}\Omega^{(2)}(4) - \frac{2}{105}\Omega^{(2)}(5)\right] ,$$

$$\sigma\left(\frac{2kT}{m}\right)^{\frac{1}{2}} J^{2}_{22} = -\left[\frac{399}{80}\Omega^{(2)}(2) - \frac{39}{14}\Omega^{(2)}(3) + \frac{127}{210}\Omega^{(2)}(4) - \frac{2}{35}\Omega^{(2)}(5) + \frac{1}{315}\Omega^{(2)}(6) + \frac{8}{315}\Omega^{(4)}(4)\right] ,$$

$$\sigma\left(\frac{2kT}{m}\right)^{\frac{1}{2}} J^{3}_{00} = -\frac{12}{5}\Omega^{(2)}(2) ,$$

$$\sigma\left(\frac{2kT}{m}\right)^{\frac{1}{2}} J^{3}_{01} = -2^{\frac{1}{2}}\left[\frac{9}{5}\Omega^{(2)}(2) - \frac{18}{35}\Omega^{(2)}(3)\right] ,$$

$$\sigma\left(\frac{2kT}{m}\right)^{\frac{1}{2}} J^{3}_{11} = -\left[\frac{41}{10}\Omega^{(2)}(2) - \frac{54}{35}\Omega^{(2)}(3) + \frac{58}{315}\Omega^{(2)}(4) + \frac{8}{189}\Omega^{(4)}(4)\right] ,$$

$$\sigma\left(\frac{2kT}{m}\right)^{\frac{1}{2}} J^{3}_{02} = -\left(\frac{11}{2}\right)^{\frac{1}{2}}\left[\frac{9}{10}\Omega^{(2)}(2) - \frac{18}{35}\Omega^{(2)}(3) + \frac{2}{35}\Omega^{(2)}(4)\right] ,$$

$$\sigma\left(\frac{2kT}{m}\right)^{\frac{1}{2}} J^{3}_{12} = -\frac{1}{11^{\frac{1}{2}}}\left[\frac{121}{8}\Omega^{(2)}(2) - \frac{1199}{140}\Omega^{(2)}(3) + \frac{187}{630}\Omega^{(2)}(4) - \frac{31}{315}\Omega^{(2)}(5) + \frac{44}{189}\Omega^{(4)}(4) - \frac{8}{189}\Omega^{(4)}(5)\right] ,$$

$$\sigma\left(\frac{2kT}{m}\right)^{\frac{1}{2}} J^{3}_{22} = -\left[\frac{3067}{480}\Omega^{(2)}(2) - \frac{121}{28}\Omega^{(2)}(3) + \frac{1931}{1980}\Omega^{(2)}(4) - \frac{31}{315}\Omega^{(2)}(5) + \frac{29}{6930}\Omega^{(2)}(6) + \frac{166}{945}\Omega^{(4)}(4) - \frac{8}{189}\Omega^{(4)}(5) + \frac{8}{2079}\Omega^{(4)}(6)\right] ,$$

$$\sigma\left(\frac{2kT}{m}\right)^{\frac{1}{2}} J^{4}_{00} = -\left[\frac{12}{5}\Omega^{(2)}(2) - \frac{4}{63}\Omega^{(2)}(3) + \frac{2}{27}\Omega^{(2)}(4)\right] .$$

For the case of power law forces we have:

$$J^0_{22} = \frac{2}{3} J^2_{00}, \quad J^0_{23} = -\frac{1}{(42)^{\frac{1}{2}}} \frac{s-4}{s} J^2_{00}, \quad J^0_{33} = \frac{31s^2-16s+16}{28s^2} J^2_{00},$$

$$J^1_{11} = \frac{2}{3} J^2_{00}, \quad J^1_{12} = -\frac{1}{3(7)^{\frac{1}{2}}} \frac{s-4}{s} J^2_{00}, \quad J^1_{22} = \frac{45s^2-16s+16}{42s^2} J^2_{00},$$

$$J^2_{00} = J^2_{00}, \quad J^2_{01} = -\frac{1}{(56)^{\frac{1}{2}}} \frac{s-4}{s} J^2_{00}, \quad J^2_{11} = \frac{205s^2-48s+48}{168s^2} J^2_{00},$$

$$J^3_{00} = \frac{3}{2} J^2_{00}.$$

For the case of elastic spheres we have:

$$J^0_{22} = \frac{2}{3} J^2_{00}, \quad J^1_{11} = \frac{2}{3} J^2_{00}, \quad J^2_{00} = -\frac{8}{5} \frac{\pi a^2}{\sigma} \left(\frac{2}{\pi}\right)^{\frac{1}{2}},$$

$$J^0_{23} = -\frac{1}{2} \left(\frac{2}{21}\right)^{\frac{1}{2}} J^2_{00}, \quad J^1_{12} = -\frac{1}{3(7)^{\frac{1}{2}}} J^2_{00}, \quad J^2_{01} = -\frac{1}{2(14)^{\frac{1}{2}}} J^2_{00},$$

$$J^0_{24} = -\frac{1}{12(21)^{\frac{1}{2}}} J^2_{00}, \quad J^1_{13} = -\frac{1}{12(42)^{\frac{1}{2}}} J^2_{00}, \quad J^2_{02} = \frac{-1}{24(14)^{\frac{1}{2}}} J^2_{00},$$

$$J^0_{33} = \frac{31}{28} J^2_{00}, \quad J^1_{22} = \frac{15}{14} J^2_{00}, \quad J^2_{11} = \frac{205}{168} J^2_{00},$$

$$J^0_{34} = -\frac{67(2)^{\frac{1}{2}}}{336} J^2_{00}, \quad J^1_{23} = -\frac{103}{336} \left(\frac{2}{3}\right)^{\frac{1}{2}} J^2_{00}, \quad J^2_{12} = -\frac{163}{672} J^2_{00},$$

$$J^0_{44} = \frac{2929}{2016} J^2_{00}, \quad J^1_{33} = \frac{5657}{4032} J^2_{00}, \quad J^2_{22} = \frac{1321}{896} J^2_{00},$$

$$J^3_{00} = \frac{3}{2} J^2_{00}, \quad J^3_{01} = -\frac{9}{28(2)^{\frac{1}{2}}} J^2_{00}, \quad J^3_{02} = -\frac{1}{112} \left(\frac{11}{2}\right)^{\frac{1}{2}} J^2_{00}$$

$$J^3_{11} = \frac{1655}{1008} J^2_{00}, \quad J^3_{12} = -\frac{4453}{4032(11)^{\frac{1}{2}}} J^2_{00}, \quad J^3_{22} = -\frac{323\,773}{177\,408} J^2_{00}.$$

APPENDIX B

Temperature dependence of Δ_1 *and* Δ_2.

T^*	Δ_1	Δ_2	T^*	Δ_1	Δ_2
0.30	+0.1544	-0.03866	2.70	+0.2336	-0.07616
0.35	+0.1264	-0.02696	2.80	+0.2430	-0.07771
0.40	+0.09098	-0.01579	2.90	+0.2505	-0.07947
0.45	+0.06134	-0.00661	3.00	+0.2555	-0.08073
0.50	+0.03331	+0.00029	3.10	+0.2638	-0.08221
0.55	+0.01320	+0.00487	3.20	+0.2680	-0.08325
0.60	+0.00185	+0.00779	3.30	+0.2745	-0.08456
0.65	-0.01154	+0.00893	3.40	+0.2793	-0.08529
0.70	-0.01935	+0.00872	3.50	+0.2844	-0.08647
0.75	-0.02089	+0.00746	3.60	+0.2903	-0.08733
0.80	-0.02066	+0.00604	3.70	+0.2940	-0.08838
0.85	-0.01846	+0.00387	3.80	+0.2988	-0.08915
0.90	-0.01448	+0.00101	3.90	+0.3024	-0.08979
0.95	-0.00891	-0.00243	4.00	+0.3065	-0.09074
1.00	-0.00169	-0.00524	4.10	+0.3092	-0.09148
1.05	+0.00450	-0.00877	4.20	+0.3122	-0.09208
1.10	+0.01203	-0.01213	4.30	+0.3153	-0.09252
1.15	+0.02096	-0.01529	4.40	+0.3183	-0.09331
1.20	+0.02799	-0.01893	4.50	+0.3222	-0.09342
1.25	+0.03908	-0.0221	4.60	+0.3245	-0.09396
1.30	+0.04544	-0.02537	4.70	+0.3274	-0.09432
1.35	+0.05556	-0.02849	4.80	+0.3292	-0.09509
1.40	+0.06678	-0.03131	4.90	+0.3323	-0.09530
1.45	+0.07250	-0.03407	5	+0.3332	-0.09591
1.50	+0.08211	-0.03680	6	+0.3499	-0.09878
1.55	+0.09243	-0.03898	7	+0.3605	-0.10054
1.60	+0.09981	-0.04219	8	+0.3682	-0.10141
1.65	+0.1047	-0.04443	9	+0.3721	-0.10230
1.70	+0.1164	-0.04692	10	+0.3758	-0.10276
1.75	+0.1256	-0.04912	20	+0.3858	-0.10368
1.80	+0.1318	-0.05092	30	+0.3865	-0.10409
1.85	+0.1384	-0.05285	40	+0.3878	-0.10385
1.90	+0.1452	-0.05479	50	+0.3872	-0.10391
1.95	+0.1524	-0.05685	60	+0.3869	-0.10429
2.00	+0.1637	-0.05870	70	+0.3879	-0.10387
2.10	+0.1604	-0.06209	80	+0.3867	-0.10399
2.20	+0.1856	-0.06438	90	+0.3860	-0.10421
2.30	+0.1960	-0.06705	100	+0.3883	-0.10408
2.40	+0.2070	-0.06947	200	+0.3882	-0.10395
2.50	+0.2187	-0.07199	300	+0.3817	-0.10402
2.60	+0.2229	-0.07406	400	+0.3888	-0.10395

$s = 12$, $\Delta_2 = -0.0986$, $\Delta_1 = +0.3633$.

APPENDIX C

The function Z.

We discuss here some of the properties of the complex function:

$$Z(z) = \pi^{-\frac{1}{2}} \int_{-\infty}^{\infty} \mathrm{d}t\, \mathrm{e}^{-t^2} \frac{1}{t-z} \tag{C.1}$$

This function is variously called the plasma dispersion function, the complex probability integral, etc. A more or less exhaustive discussion of its mathematical properties is given by J. B. Rosser, *Theory and Application of*

$$\int_0^z \mathrm{e}^{-x^2} \mathrm{d}x \quad \textit{and} \quad \int_0^z \mathrm{e}^{-p^2 y^2} \mathrm{d}y \int_0^y \mathrm{e}^{-x^2} \mathrm{d}x$$

(Mapleton House Publishers, Brooklyn, N. Y., 1948). The most extensive numerical tables of $Z(z)$ are given by B. D. Fried and S. D. Conte, *The Plasma Dispersion Function* (Academic Press, New York, 1961).

From the integral representation it is clear that Z is an analytic function of z except for points on the real axis, which is a branch line or line of discontinuity for Z. The function in the upper and lower half planes is related by:

$$Z(z) = -Z(-z) \ . \tag{C.2}$$

If we form the derivative with respect to z under the integral in (C.1) and then integrate by parts, we obtain the following differential equation for Z:

$$Z' = -2(1 + zZ) \ . \tag{C.3}$$

The behaviour of Z along the real axis is given by

$$\lim_{\epsilon \to 0^+} Z(x + i\epsilon) = Y(x) + i\pi^{\frac{1}{2}} \mathrm{e}^{-x^2} \ , \tag{C.4}$$

where

$$Y(x) = \pi^{-\frac{1}{2}} \mathrm{P} \int_{-\infty}^{\infty} \mathrm{d}t\, \mathrm{e}^{-t^2} \frac{1}{t-x} \ , \tag{C.5}$$

in which P stands for the principal value of the integral. For $Y(x)$ we have the simple integral representation:

$$Y(x) = -2\, e^{-x^2} \int_0^x dt\, e^{t^2} \, . \qquad (C.6)$$

The simplest way to get the results (C.4) and (C.6) is to note that the general solution of the differential equation (C.3) is

$$Z(z) = Y(z) + a\, e^{-z^2} \, , \qquad (C.7)$$

where a is a constant which is determined by the condition:

$$\lim_{\epsilon \to 0^+} Z(i\epsilon) = i\, \pi^{\frac{1}{2}} \, . \qquad (C.8)$$

This last condition is obtained from the integral representation (C.1) when we note that

$$\lim_{\epsilon \to 0^+} \frac{1}{t - i\epsilon} = \mathrm{P}\, \frac{1}{t} + i\pi\, \delta(t) \, , \qquad (C.9)$$

where P denotes the principal value and $\delta(t)$ is the Dirac delta-function. Using (C.2) we readily see that

$$\lim_{\epsilon \to 0^+} Z(x - i\epsilon) = Y(x) - i\pi^{\frac{1}{2}}\, e^{-x^2} \, . \qquad (C.10)$$

If we expand the integrand of (C.1) in inverse powers of z and then integrate term by term, we obtain the following *asymptotic* expansion for large $|z|$:

$$Z(z) \sim - \sum_{n=0}^{\infty} \frac{\Gamma(n+\frac{1}{2})}{\Gamma(\frac{1}{2})}\, z^{-2n-1} \, . \qquad (C.11)$$

By repeated differentiation of the differential equation (C.3) we can show that the n'th derivative of Z may be expressed in the form:

$$Z^{(n)}(z) = (-)^n \left[2\, G_n(z) + H_n(z)\, Z(z) \right] \, , \qquad (C.12)$$

where $G_n(z)$ and $H_n(z)$ are polynomials which satisfy the recursion relations:

$$H_{n+1} = -H_n' + 2z H_n \ ,$$
$$G_{n+1} = -G_n' + H_n \ , \qquad (C.13)$$

and

$$H_{n+1} - 2z H_n + 2n H_{n-1} = 0 \ ,$$
$$G_{n+1} - 2z G_n + 2n G_{n-1} = 0 \ . \qquad (C.14)$$

Thus H_n and G_n satisfy the same two-term recursion relation, but differ in their initial conditions:

$$H_0 = 1 \ , \quad H_1 = 2z \ , \quad G_0 = 0 \ , \quad G_1 = 1 \ . \qquad (C.15)$$

We see immediately that $H_n(z)$ is the Hermite polynomial:

$$H_n(z) \equiv (-)^n \, e^{z^2} \frac{d^n}{dz^n} e^{-z^2} \ . \qquad (C.16)$$

(See, e.g., *Higher Transcendental Functions*, Erdelyi et al., (Mc Graw Hill Publishing Co., New York, 1964) Vol. II, chapter X. The first few of the polynomials H_n and G_n are given in a table at the end of this appendix. A list of closely related Polynomials through $n = 16$ is given by Rosser, *op. cit.*, p. 47.

The importance of these polynomials stems from the theorem that if we put

$$Z_n(z) \equiv -2 \frac{G_n(z)}{H_n(z)} \ , \qquad (C.17)$$

then

$$\lim_{n \to \infty} Z_n(z) = Z(z) \ , \qquad (C.18)$$

everywhere except on the real z axis. This theorem follows when we note that $Z_n(z)$ is the n'th approximant of the continued fraction:

$$Z(z) = \cfrac{-2}{2z - \cfrac{2}{2z - \cfrac{4}{2z - \cfrac{6}{2z - \cfrac{8}{\cdots}}}}} \qquad (C.19)$$

which converges for z not on the real axis. See, e.g., H. S. Wall, *Continued Fractions*, (D. Van Nostrand Company, New York, 1948), esp. chapter XVIII.

Table 1
The first eight polynomials H_n and G_n

$H_0 = 1$	$G_0 = 0$
$H_1 = 2z$	$G_1 = 1$
$H_2 = 2(2z^2 - 1)$	$G_2 = 2z$
$H_3 = 4z(2z^2 - 3)$	$G_3 = 4(z^2 - 1)$
$H_4 = 4(4z^4 - 12z^2 + 3)$	$G_4 = 4z(2z^2 - 5)$
$H_5 = 8z\ (4z^4 - 20z^2 + 15)$	$G_5 = 8(2z^4 - 9z^2 + 4)$
$H_6 = 8(8z^6 - 60z^4 + 90z^2 - 15)$	$G_6 = 8z(4z^4 - 28z^2 + 33)$
$H_7 = 16z\ (8z^6 - 84z^4 + 210z^2 - 105)$	$G_7 = 16(4z^6 - 40z^4 + 87z^2 - 24)$
$H_8 = 16\ (16z^8 - 224z^6 + 840^4 - 840z^2 + 105)$	$G_8 = 16z(8z^6 - 108z^4 + 370z^2 - 279)$

SUBJECT INDEX